W9-DJD-667

Unit III

Science Projects

Phil Schlemmer

Illustrated by Patricia A. Sussman

LEARNING ON YOUR OWN!

Individual, Group, and Classroom
Research Projects for
Gifted and Motivated Students

The Center for Applied Research in Education, Inc.
West Nyack, New York

10   9   8   7   6   5   4

Library of Congress Cataloging-in-Publication Data

Schlemmer, Phillip L.
    Science projects.

    (Learning on your own: unit 3)
    1. Science—Study and teaching. 2. Science—
Experiments. 3. Research—Methodology. 4. Creative
thinking (Education) 5. Gifted children—Education—
Science. 6. Independent study. I. Title. II. Series:
Schlemmer, Phillip L. Learning on your own! ; unit 3.
Q181.S34   1987   372.3'5   86-20734
ISBN 0-87628-509-4

PRINTED IN THE UNITED STATES OF AMERICA

Dedication

This book is dedicated to my wife, Dori. Without her unending support, tireless editorial efforts, thoughtful criticisms, and patience, I could not have finished my work. Thank you, Dori.

Acknowledgments

My collaborator and co-teacher for eight years, Dennis Kretschman, deserves special mention at this juncture. Together we developed the activities, projects, and courses that became a "learning to learn" curriculum. Dennis designed and taught several of the projects described in these pages, and he added constantly to the spirit and excitement of an independent learning philosophy that gradually evolved into this set of five books. I deeply appreciate the contribution Dennis has made to my work.

I would also like to thank the following people for their advice, support, and advocacy: J. Q. Adams; Dr. Robert Barr; Robert Cole, Jr.; Mary Dalheim; Dr. John Feldhusen; David Humphrey; Bruce Ottenweller; Dr. William Parrett; Ed Saunders; Charles Whaley; and a special thanks to all the kids who have attended John Ball Zoo School since I started working on this project: 1973–1985.

About the Author

PHIL SCHLEMMER, M.Ed., has been creating and teaching independent learning projects since 1973, when he began his master's program in alternative education at Indiana University. Assigned to Grand Rapids, Michigan, for his internship, he helped develop a full-time school for 52 motivated sixth graders. The school was located at the city zoo and immediately became known as the "Zoo School." This program became an experimental site where he remained through the 1984–85 school year, with one year out as director of a high school independent study program.

Presently working as a private consultant, Mr. Schlemmer has been presenting in-services and workshops to teachers, parents, administrators, and students for more than 13 years and has published articles in *Phi Delta Kappan* and *Instructor.*

Foreword

This series of books will become invaluable aids in programs for motivated, gifted, and talented children. They provide clear guidelines and procedures for involving these children in significant learning experiences in research and high level thinking skills while not neglecting challenging learning within the respective basic disciplines of science, mathematics, social studies, and writing. The approach is one that engages the interests of children at a deep level. I have seen Phil Schlemmer at work teaching with the materials and methods presented in these books and have been highly impressed with the quality of learning which was taking place. While I recognized Phil is an excellent teacher, it nevertheless seemed clear that the method and the materials were making a strong and significant contribution to the children's learning.

Children will learn how to carry out research and will become independent lifelong learners through the skills acquired from the program of studies presented in these books. Success in independent study and research and effective use of libraries and other information resources are not simply products of trial-and-error activity in school. They are products of teacher guidance and stimulation along with instructional materials and methods and an overall system which provides the requisite skills and attitudes.

All of the material presented in this series of books has undergone extensive tryout. The author has also spent thousands of hours developing, writing, revising, and editing, but above all he has spent his time conceptualizing and designing a dynamic system for educating motivated, gifted, and talented youth. The net result is a program of studies which should make an invaluable contribution to the education of these youth. And, above all that, I am sure that if it is taught well, the kids will love it.

John F. Feldhusen, Ph.D., Director
Gifted Education Resource Institute
Purdue University
West Lafayette, Indiana 47907

About Learning on Your Own!

In the summer of 1973, I was offered the opportunity of a lifetime. The school board in Grand Rapids, Michigan authorized a full-time experimental program for 52 motivated sixth-grade children, and I was asked to help start it. The school was described as an environmental studies program, and its home was established in two doublewide house trailers that were connected and converted into classrooms. This building was placed in the parking lot of Grand Rapids' municipal zoo (John Ball Zoological Gardens). Naturally, the school came to be known as "The Zoo School."

The mandate for the Zoo School staff was clear—to build a challenging, stimulating, and interesting curriculum that was in no way limited by the school system's stated sixth-grade objectives. Operating with virtually no textbooks or "regular" instructional materials, we had the freedom to develop our own projects and courses, schedule our own activities, and design our own curriculum.

Over a period of ten years, hundreds of activities were created to use with motivated learners. This was a golden opportunity because few teachers are given a chance to experiment with curriculum in an isolated setting with the blessing of the school board. When a project worked, I wrote about it, recorded the procedures that were successful, filed the handouts, and organized the materials so that someone else could teach it. The accumulation of projects for motivated children led to a book proposal which, in turn, led to this five-book series. *Learning on Your Own!* is based entirely on actual classroom experience. Every project and activity has been used successfully with children in the areas of

- Research Skills
- Writing
- Science
- Mathematics
- Social Studies

As the books evolved and materialized over the years, it seemed that they would be useful to classroom teachers, especially in the upper and elementary and junior high grades. This became increasingly clear as teachers from a wide variety of settings were presented with ideas from the books. Teachers saw different uses for the projects, based upon the abilities of their students and their own curricular needs.

Learning on Your Own! will be useful to you for any of the following reasons:

- If a curricular goal is to teach children to be independent learners, then skill development is necessary. The projects in each book are arranged according to the level of independence that is required—the early projects can be used to *teach* skills; the later ones require their *use*.

- These projects prepare the way for students confidently to make use of higher-level thinking skills.

- A broad range of students can benefit from projects that are skill-oriented. They need not be gifted/talented.

- On the other hand, teachers of the gifted/talented will see that the emphasis on independence and higher-level thinking makes the projects fit smoothly into their curricular goals.

- The projects are designed for use by one teacher with a class of up to 30 students. They are intentionally built to accommodate the "regular" classroom teacher. Projects that require 1-to-1 or even 1-to-15 teacher-student ratios are of little use to most teachers.

- The books do not represent a curriculum that must be followed. Gifted/talented programs may have curricula based upon the five-book series, and individual situations may allow for the development of a "learning to learn" curriculum. Generally speaking, however, each project is self-contained and need not be a part of a year-long progression of courses and projects.

- Each project offers a format that can be used even if the *content* is changed. You may, with some modification, apply many projects toward subject material that is already being taught. This provides a means of delivering the same message in a different way.

- Most teachers have students in their classes capable of pursuing projects that are beyond the scope of the class as a whole. These books can be used to provide special projects for such students so that they may learn on their own.

- One of the most pervasive concepts in *Learning on Your Own!* is termed "kids teaching kids." Because of the emphasis placed on students teaching one another, oral presentations are required for many projects. This reinforces the important idea that not only can students *learn,* they can also *teach.* Emphasis on oral presentation can be reduced if time constraints demand it.

- The premise of this series is that children, particularly those who are motivated to learn, need a base from which to expand their educational horizons. Specifically, this base consists of five important components of independent learning:

—skills

—confidence

—a mandate to pursue independence

—projects that show students *how* to learn on their own

—an opportunity to practice independent learning

Learning on Your Own! places primary emphasis on the motivated learner, the definition of which is left intentionally ambiguous. It is meant to include most normal children who have natural curiosities and who understand the need for a good education. Motivated children are important people who deserve recognition for their ability and desire to achieve. The trend toward understanding the special needs and incredible potential of children who enjoy the adventure and challenge of learning is encouraging. Teachers, parents, business people, community leaders, and concerned citizens are beginning to seriously ask, "What can we do for these young people who want to learn?"

Creating a special program or developing a new curriculum is not necessarily the answer. Many of the needs of these children can be met in the regular classroom by teaching basic independent learning skills. No teacher can possibly master and teach all of the areas that his or her students may be interested in studying, but every teacher has opportunities to place emphasis on basic learning skills. A surprising number of children become more motivated as they gain skills that allow them to learn independently. "Learning on your own" is an important concept because it, in itself, provides motivation. You can contribute to your students' motivation by emphasizing self-confidence and skill development. One simple project during a semester can give students insight into the usefulness of independent learning. One lesson that emphasizes a skill can bring students a step closer to choosing topics, finding information, planning projects, and making final presentations without assistance. By teaching motivated students *how to learn on their own,* you give them the ability to challenge themselves, to transcend the six-hour school day.

Beyond meeting the immediate needs of individual students, teaching children how to learn on their own will have an impact on their adult lives and may affect society itself. It is easy to discuss the day-to-day importance of independent learning in one breath, and in the next be talking of the needs of adults 30 years from now. This five-book series is based upon the assumption that educating children to be independent learners makes sense in a complicated, rapidly changing, unpredictable world. Preparing today's children for tomorrow's challenges is of paramount importance to educators and parents, but a monumental task lies in deciding what can be taught that will have lasting value in years to come. What will people need to know in the year 2001 and beyond? Can we accurately prescribe a set of facts and information that will be *necessary* to an average citizen 10, 20, or 30 years from now? Can we feel confident that what we teach will be useful, or even relevant, by the time our students become adults? Teaching children to be independent learners is a compelling response to these difficult, thought-provoking questions.

How to Use Learning on Your Own!

Learning on Your Own! can be used in many ways. The projects and the overall design of the books lend themselves to a variety of applications, such as basic skill activities, full-class units or courses, small-group projects, independent study, and even curriculum development. Regardless of how the series is to be implemented, it is important to understand its organization and recognize what it provides. Like a good cookbook, this series supplies more than a list of ingredients. It offers suggestions, advice, and hints; provides organization and structure; and gives time lines, handouts, and materials lists. In other words, it supplies everything necessary for you to conduct the projects.

These books were produced with you in mind. Every project is divided into three general sections to provide uniformity throughout the series and to give each component a standard placement in the material. The first section, Teacher Preview, gives a brief overview of the scope and focus of the project. The second section, Lesson Plans and Notes, outlines a detailed, hour-by-hour description. After reading this, every nuance of the project should be understood. The third section, Instructional Materials, supplies the "nuts-and-bolts" of the project— reproducible assignment sheets, instructional handouts, tests, answer sheets, and evaluations.

Here is a concise explanation of each of the three sections. Read this material before going further to better understand how the projects can be used.

Teacher Preview

The Teacher Preview is a quick explanation of what a project accomplishes or teaches. It is divided into seven areas, each of which provides specific information about the project:

Length of Project: The length of each project is given in classroom hours. It does not take into account homework or teacher-preparation time.

Level of Independence: Each project is identified as "basic," "intermediate," or "advanced" in terms of how much independence is required of students. The level of independence is based primarily on how many decisions a student must make and how much responsibility is required. It is suggested that students who have not acquired independent learning skills, regardless of their grade level, be carefully introduced to advanced projects.

For teachers who are interested, there is a correlation between the skill development mentioned here and the progression to higher-level thinking skills typified by Benjamin Bloom's "Taxonomy of Educational Objectives":

Level of Independence	*Bloom's Taxonomy*
Basic	Knowledge
	Comprehension
Intermediate	Application
	Analysis
Advanced	Synthesis
	Evaluation

Goals: These are straightforward statements of what a project is designed to accomplish. Goals that recur throughout the series deal with skill development, independent learning, and "kids teaching kids."

During This Project Students Will: This is a list of concise project objectives. Occasionally, some of these statements become activities rather than objectives, but they are included because they help specify what students will do during the course of a project.

Skills: Each project emphasizes a specific set of skills, which are listed in this section. Further information about the skills is provided in the "Skills Chart." You may change the skill emphasis of a project according to curricular demands or the needs of the students.

Handouts Provided: The handouts provided with a project are listed by name. This includes assignment sheets, informational handouts, tests, and evaluation forms.

Project Calendar: This is a chart that graphically shows each hour of instruction. Since it does not necessarily represent consecutive days, lines are provided for you to pencil in dates. The calendar offers a synopsis of each hour's activity and also brief notes to clue you about things that must be done:

PREPARATION REQUIRED	STUDENTS TURN IN WORK
NEED SPECIAL MATERIALS	RETURN STUDENT WORK
HANDOUT PROVIDED	ANSWER SHEET PROVIDED

Lesson Plans and Notes

The lesson plan is a detailed hour-by-hour description of a project, explaining its organization and presentation methods. Projects can be shortened by reducing the time spent on such things as topic selection, research, and presentation; however, this necessitates de-emphasizing skills that make real independent study possible. Alternately, a project may require additional hours if students are weak in particular skill areas or if certain concepts are not thoroughly understood.

Each hour's lesson plan is accompanied by notes about the project. Some notes are fairly extensive if they are needed to clarify subject matter or describe a process.

Instructional Material

There are five types of reproducible instructional materials included in *Learning on Your Own!* Most projects can be run successfully with just a Student Assignment Sheet; the rest of the materials are to be used as aids at your discretion.

Student Assignment Sheets: Virtually every project has an assignment sheet that explains the project and outlines requirements.

Additional Handouts: Some projects offer other handouts to supply basic information or provide a place to record answers or research data.

Tests and Quizzes: Tests and quizzes are included with projects that present specific content. Since most projects are individualized, the activities themselves are designed to test student comprehension and skill development.

Evaluation Sheets: Many projects provide their own evaluation sheets. In addition, the Teacher's Introduction to the Student Research Guide (see the Appendix) contains evaluations for notecards, posters, and oral presentations. Some projects also supply self-evaluation forms so that students can evaluate their own work.

Forms, Charts, Lists: These aids are provided throughout the series. They are designed for specific situations in individual projects.

OTHER FEATURES OF
LEARNING ON YOUR OWN!

In addition to the projects, each book in the series offers several other useful features:

Grade Level: A grade level notation of upper elementary, junior high, and/or high is shown next to each project in the table of contents. Because this series was developed with gifted/talented/motivated sixth graders, junior high is the logical grade level for most projects; thus, generally speaking, these projects are most appropriate for students in grades 6–8.

Skills Chart: This is a chart listing specific independent learning skills that may be applied to each project. It is fully explained in its introductory material.

Teacher's Introduction to the Student Research Guide: This introduction is found in the Appendix. It offers suggestions for conducting research projects and provides several evaluation forms.

Student Research Guide: Also found in the Appendix, this is a series of checklists that can be used by students working on individualized projects. The Daily Log, for example, is a means of having students keep track of their own progress. In addition to the checklists, there are instructional handouts on basic research skills.

General Notes

Examine the *structure* of the projects in each book, even if the titles do not fit specific needs. Many projects are so skill-oriented that content can be drastically altered without affecting the goals.

Many projects are dependent upon resource materials; the more sources of information, the better. Some way of providing materials for the classroom are to

- Ask parents for donations of books and magazines.
- Advertise for specific materials in the classified section of the newspaper.
- Check out library materials for a mini-library in the classroom.
- Gradually purchase useful materials as they are discovered.
- Take trips to public libraries and make use of school libraries.

Students may not initially recognize the value of using notecards. They will soon learn, however, that the cards allow data to be recorded as individual facts that can be arranged and rearranged at will.

"Listening" is included as an important skill in most projects. In lecture situations, class discussions, and when students are giving presentations, you should require students to listen and respect the right of others to listen.

Provide time for grading and returning materials to students during the course of a project. The Project Calendar is convenient for planning a schedule.

A visual display is often a requirement for projects in this series. Students usually choose to make a poster, but there are other possibilities:

mural	collage	demonstration	dramatization
mobile	model	display or exhibit	book, magazine, or pamphlet
diorama	puppet show	slide show	

When students work on their own, your role changes from information supplier to learning facilitator. It is also important to help students solve their own problems so that momentum and forward progress are maintained.

A FINAL NOTE FROM THE AUTHOR

Learning on Your Own! provides the materials and the structure that are necessary for individualized learning. The only missing elements are the enthusiasm, vitality, and creative energy that are needed to ignite a group of students and set them diligently to work on projects that require concentration and perseverance. I hope that *my* work will make *your* work easier by letting you put your efforts into quality and innovation. The ability to learn independently is perhaps the greatest gift that can be conferred upon students. Give it with the knowledge that it is valuable beyond price, uniquely suited to each individual, and good for a lifetime!

Phil Schlemmer

About This Book

The projects in this book are designed to make a variety of science topics available to the students in your classroom. Some of them offer basic information and others emphasize independent learning. The ultimate goal of each project is to help students realize that they are *capable* of learning about scientific topics, and that such learning can often be done independently. In today's world the term "science" covers almost every aspect of life. There is no conceivable way that current areas of specialization can be comprehensively presented in school, let alone trying to keep up with constantly emerging new fields of study. It makes sense, therefore, to place emphasis on independent learning when guiding students through the maze of interrelating subjects and facts that comprise the study of science.

Science Projects begins with a group of projects and activities that teach basic information about physics, chemistry, and vertebrate animals. An extensive amount of teacher support material and background information is supplied. The primary goal of these projects is to show students that they *can* learn about such intimidating subjects as physics, chemistry, and zoology. The remaining projects are designed to emphasize research and independent learning. Many of these projects are about the animal kingdom, but there are also projects about plant science, astronomy, and geology, among others. It is very important to realize the value of the project formats that are presented. Using the structure and format of Oceanography, for example, you could construct new projects titled Desert Life, Jungles, or Birds.

Teachers in the upper elementary and junior high grades often feel science is one of the most difficult subjects to teach with confidence. For a number of reasons, they frequently find that teaching such subjects as astronomy, zoology, chemistry, physics, botany, or geology is difficult. This is because these sciences are diverse, interdependent, technical, time-consuming, and demanding. Further, the availability of good curriculum materials is often limited. While it is true that teaching science can be difficult, and perhaps intimidating, it is not impossible. More important, science is a *necessary* part of an education that must prepare students for the next century. The teacher who can lay a simple, basic scientific groundwork for students in their upper elementary and junior high years is providing an extremely valuable educational service. This book should be a useful tool for such a teacher.

THE SKILLS CHART

Science Projects is based upon skill development. The projects are arranged according to the amount of independence required, and a list of skills is provided for every project in the book. A comprehensive Skills Chart is included here to help define the skill requirements of each project. Many of them are basic, common-sense skills that are already being taught in your classes.

The Skills Chart is divided into seven general skill areas: research, writing, planning, problem solving, self-discipline, self-evaluation, and presentation. Reading is not included on the chart because it is assumed that reading skills will be used with virtually every project.

The key tells if a skill is prerequisite (#), primary (X), secondary (0), or optional (*) for each project in the book. These designations are based upon the way the projects were originally taught; you may want to shift the skill emphasis of a project to fit the needs of your particular group of students. It is entirely up to you to decide how to present a project and what skills to emphasize. The Skills Chart is only a guide.

Examination of the chart quickly shows which skills are important to a project and which ones may be of secondary value. A project may be changed or rearranged to redirect its skill requirements. The projects in this book are designed to *teach* the use of skills. If a project's Teacher Preview lists twenty skills, but you want to emphasize only three or four of them, that is a perfectly legitimate use of the project.

Evaluating students on their mastery of skills often involves subjective judgments; each student should be evaluated according to his or her *improvement* rather than by comparison with others. Several projects supply evaluation forms to help with this process. In addition, the Teacher's Introduction to the Student Research Guide provides evaluations for notecards, posters, and oral presentations.

A blank Skills Chart is included at the end of the Student Research Guide in the Appendix. This chart can be helpful in several ways:

• Students can chart their own skill progression through a year. Give them a chart and tell them to record the title of a project on the first line. Have them mark the skills *you* have decided to emphasize with the project. This way, students will see *exactly* what skills are being taught and which ones they are expected to know how to use. As projects are continued through the year, the charts will indicate skill development.

• Use the chart to organize the skill emphasis of projects that did not come from this book. Quite often, projects have the potential to teach skills, but they are not organized to do so. An entire course or even a curriculum can be organized according to the skill development on the chart.

• The Skills Chart can be used as a method of reporting to parents. By recording the projects and activities undertaken during a marking period in the left-hand column, a mark for each of the 48 skills can be given. For example, a number system can be used:

1—excellent
2—very good
3—good

4—fair

5—poor

• A simpler method of reporting to parents is to give them a copy of the Skills Chart without marks and use it as the basis for a discussion about skill development.

Finally, most teachers have little or no experience teaching some of the skills listed on the chart. There is plenty of room for experimentation in the field of independent learning, and there are no established "correct" methods of teaching such concepts as problem solving, self-evaluation, and self-discipline. These are things that *can* be taught, but your own teaching style and philosophy will dictate how you choose to do it. The skills listed on this chart should be recognizable as skills that are worth teaching, even if you have not previously emphasized them.

SKILLS CHART: SCIENCE

*Prerequisite Skills* Students must have command of these skills.

X *Primary Skills* Students will learn to use these skills; they are necessary to the project.

0 *Secondary Skills* These skills may play an important role in certain cases.

* *Optional Skills* These skills may be emphasized but are not required.

	RESEARCH									WRITING						PLANNING				
	PREPARING BIBLIOGRAPHIES	COLLECTING DATA	INTERVIEWING	WRITING LETTERS	LIBRARY SKILLS	LISTENING	MAKING NOTECARDS	OBSERVING	SUMMARIZING	GRAMMAR	HANDWRITING	NEATNESS	PARAGRAPHS	SENTENCES	SPELLING	GROUP PLANNING	ORGANIZING	OUTLINING	SETTING OBJECTIVES	SELECTING TOPICS
INCLINED PLANES		X				X		X	0	0	X						X			
LEVERS		X				X		X	0	0	X						X			
PULLEYS		X				X		X	0	0	X						X			
ENERGY & ELECTRICITY		X				X		X	0	0	X						X			
SPEED OF TOBOGGANS		X				X		X	0	0	X					X	X			
CHEMISTRY	*	X	*		*	X		X	0	0	X				X		X			
ANIMAL TAXONOMY	*	X	*	*	*	X	*	X	X	X	X				X		X	X		0
VERTEBRATE ANIMALS	X	X	*	*	X	X	X	0	X	X	X	X	X	X	X	*	X	0	X	X
ENTOMOLOGY	X	X	*	*	X	X	X	0	X	X	X	X	X	X	X	*	X	0	X	X
OCEANOGRAPHY	X	X	*	*	X	X	X	0	X	X	X	X	X	X	X	*	X	0	X	X
ANIMALS IN THE WILD	X	X	0	0	X	X	0	0	X	X	X	X	X	X	X	*	X	X	X	X
IMAGINARY ANIMALS	X	X	0	0	X	X	0	0	X	X	X	X	X	X	X	*	X	X	X	X
NATURAL ENVIRONMENTS	X	X	0	0	X	X	X	0	X	X	X	X	X	X	X	X	X	0	X	X
INDEPENDENT PROJECTS	#	X	0	0	#	X	#	X	X	X	X	X	#	#	X	0	X	X	X	X
SCIENCE FAIR	#	#	0	0	#	0	#	X	X	X	X	X	#	#	#	0	X	X	X	X

© 1987 by The Center for Applied Research in Education, Inc.

SKILLS CHART: SCIENCE

Column groups: **PROBLEM SOLVING** (Basic Mathematics Skills — Working w/Limited Resources) · **SELF-DISCIPLINE** (Accepting Responsibility — Working with Others) · **SELF-EVALUATION** (Personal Motivation — Setting Personal Goals) · **PRESENTATION** (Creative Expression — Writing)

Basic Mathematics Skills	Diverge-Converge-Evaluate	Following & Changing Plans	Identifying Problems	Meeting Deadlines	Working w/Limited Resources	Accepting Responsibility	Concentration	Controlling Behavior	Following Project Outlines	Individualized Study Habits	Persistence	Sharing Space	Taking Care of Materials	Time Management	Working with Others	Personal Motivation	Self-Awareness	Sense of "Quality"	Setting Personal Goals	Creative Expression	Creating Strategies	Diorama & Model Building	Drawing/Sketching/Graphing	Poster Making	Public Speaking	Self-Confidence	Teaching Others	Writing
X	X		X		0		X	0		X	X	0	0		0	X		X					0					
X	X		X		0		X	0		X	X	0	0		0	X		X					0					
X	X		X		0		X	0		X	X	0	0		0	X		X					0					
X			X		0		X	0		X	X	0	0		0	X		X					0					
X		X		0	0	X	X	X	X	X	X	X	0	X	X	X	X	X					X					
X		X		0	0	0	X	X	0	X	X	0	0		0	X	X	X	X				X	0	0	X	0	
							X																					X
	X	X	X	X	X	X	X	X	X	X	X	X	X	X	*	X	X	X	X	X	X	0	X	X	X	X	X	X
	X	X	X	X	X	X	X	X	X	X	X	X	X	X	*	X	X	X	X	X	X	0	X	X	X	X	X	X
	X	X	X	X	X	X	X	X	X	X	X	X	X	X	*	X	X	X	X	X	X	0	X	X	X	X	X	X
	X	X	X	X	X	X	X	X	X	X	X	X	X	X	*	X	X	X	X	X	X	X	X	X	*	X	X	X
	X	X	X	X	X	X	X	X	X	X	X	X	X	X	*	X	X	X	X	X	X	X	X	X	*	X	X	X
	X	X	X	X	X	X	X	X	X	X	X	X	X	X	X	X	X	X	X	X	X	*	X	X	*	X	X	X
	X	X	X	X	X	#	X	#	#	#	X	X	X	X	0	#	X	X	X	X	#	X	X	#	#	X	X	X
0	X	X	X	X	X	#	X	#	#	#	X	X	X	X	0	#	X	X	X	X	#	X	X	#	#	X	X	X

Contents

Science Projects   **Grade Level**   **Page**

PHYSICS PROJECTS

Teacher Preview

Project Topics: Inclined Planes
Levers
Pulleys
Length of Project: 7 hours
Level of Independence: Basic
Goals:

1. To explain what inclined planes, levers, and pulleys are.
2. To provide "worksheet" projects to go along with the regular science textbook.
3. To provide an opportunity for students to see how certain science problems are solved.
4. To incorporate mathematics into a science project.

During This Project Students Will:

1. Demonstrate their understanding of the terms "load," "work," and "force."
2. Use their mathematics skills by working with equations and related equations.
3. Solve simple inclined plane, lever, and pulley problems.

Skills:

Listening	Individualized study habits
Observing	Persistence
Neatness	Personal motivation
Organizing	Sense of "quality"
Divergent-convergent-evaluative thinking	Basic mathematics skills
Identifying problems	Collecting data
	Concentration

Handouts Provided:

- "Terms and Symbols Reference Sheet"
- "Information Sheet" for each area of study
- "Problem Sheet" for each area of study

PROJECT CALENDAR:

HOUR 1: _____ Introduction to the study of physics. Explanation of the terms "load," "work," and "force." Three simple machines are also introduced: inclined plane, lever, and pulley. HANDOUT PROVIDED PREPARATION REQUIRED	**HOUR 2:** _____ Discussion of inclined planes, followed by problem solving on the board. Optional classroom experiments for data collection can be done. HANDOUT PROVIDED	**HOUR 3:** _____ Students complete the "Inclined Planes Problem Sheet," and answers are discussed. HANDOUT PROVIDED ANSWER SHEET PROVIDED PREPARATION REQUIRED
HOUR 4: _____ Discussion of levers, followed by problem solving on the board. Optional classroom experiments for data collection can be done. HANDOUT PROVIDED	**HOUR 5:** _____ Students complete the "Levers Problem Sheet," and answers are discussed. HANDOUT PROVIDED ANSWER SHEET PROVIDED PREPARATION REQUIRED	**HOUR 6:** _____ Discussion of pulleys, followed by problem solving on the board. Optional classroom experiments for data collection can be done. HANDOUT PROVIDED ANSWER SHEET PROVIDED
HOUR 7: _____ Students complete the "Pulleys Problem Sheet," and answers are discussed. HANDOUT PROVIDED ANSWER SHEET PROVIDED PREPARATION REQUIRED	**HOUR 8:** _____	**HOUR 9:** _____

Lesson Plans and Notes

INCLINED PLANES (Hours 1, 2, and 3)

HOUR 1: Introduce students to the study of physics and hand out the "Terms and Symbols Reference Sheet." Discuss the terms "work," "load," and "force" and present examples of simple machines (inclined plane, lever, pulleys). Point out that these machines are used in everyday life, not to reduce the amount of *work* that must be done, but to reduce the amount of *force* needed to get the work done.

HOUR 2: Give students the "Information Sheet" for inclined planes. Spend the hour doing problems from the handout on the board. As an optional activity, data can be obtained experimentally in class with an inclined plane and a calibrated spring scale. Remember to account for friction if this is done. You may also want to create several additional problems for students to work on at their desks.

Note:

- Friction must be explained as a factor in the actual amount of force needed to move an object with a lever, a pulley, or an inclined plane. The equations in this project do *not* take friction into account, so the answers are *ideal*.

HOUR 3: Give students the "Problem Sheet" for inclined planes to work on in class. Answers are provided on the "Teacher's Answer Key." An extra-credit question should be developed for students who finish early.

LEVERS (Hours 4 and 5)

HOUR 4: Give students the "Information Sheet" for levers. Spend the hour doing problems from the handout on the board. As an optional activity, data can be obtained experimentally in class with a calibrated lever, a sharp-edged fulcrum, and a set of accurate balance weights. Other experiments can be set up that use the lever (a teeter-totter experiment, for example). You may also want to create several additional problems for students to work on at their desks.

Note:

- For lever problems, d_L (load distance) and d_F (force distance) are pure numbers, without units, because the equation for force depends only on a ratio between them. In other words, the units cancel out in the equation for force. Additionally, the units of distance can be any measure of length and the fraction $\frac{d_L}{d_F}$ will remain the same, so it does not matter if the numbers have units of centimeters, inches, feet, or none at all. A stick divided into any number of equal, arbitrary lengths can be used to calculate the answers to lever problems. Be sure to explain this to your students since no units are given for d_L and d_F in lever problems.

HOUR 5: Give students the "Problem Sheet" for levers to work on in class. Answers are provided on the "Teacher's Answer Key." An extra-credit question should be developed for students who finish early.

PULLEYS (Hours 6 and 7)

HOUR 6: Give students the "Information Sheet" for pulleys. Note that answers to the problems on this sheet are provided on the "Teacher's Answer Key." Spend the hour doing problems from the handout on the board. As an optional activity, data can be obtained experimentally in class with various combinations of pulleys and objects of known mass (weight).

Notes:

- For pulley problems, the lines attached to the load represent force vectors lifting, or moving the load. Each line reduces the amount of force needed to lift the load, and thus the equation:

$$F = \frac{\text{LOAD}}{\text{number of support lines}}$$

This means the load weight divided by the number of lines exerting an *upward* force on the load:

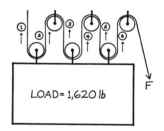

There are six support lines on this load, so the force equation is

$$F = \frac{1620 \text{ lb}}{6}$$

$$F = 270 \text{ lb}$$

A force of 270 lb must be exerted on the end line to lift a load of 1,620 lb. This means that each support line is exerting an upward force of 270 lb on the load; therefore, each support line represents a vector force of 270 lb.

- Vector forces may be explained to the class, but this is not necessary to understand the handouts.

HOUR 7: Give students the "Problem Sheet" for pulleys to work on in class. Answers are provided on the "Teacher's Answer Key." An extra credit question should be developed for students who finish early.

General Notes About This Project:

- Require students to show each step in their problem solutions. The format that is used on the "Teacher's Answer Key" is an effective problem-solving procedure that can be used.
- It is up to you to determine how extensively students will get into problem solving for inclined planes, levers, and pulleys. Additional problems, to supplement the ones provided with this project, can easily be developed by changing the numbers in handout questions.

- Students should be able to manipulate a simple equation to create new, related equations. Specifically, if $W = F \times D$, students should be able to deduce two new equations: $F = W \div D$ and $D = W \div F$.
- A final test could be developed covering the basic concepts taught about simple machines. This has not been provided, since the questions should be based on how well your students understand the physical and mathematical concepts involved.
- For optional classroom experiments which are very helpful in presenting this project, you need: an inclined plane, a calibrated spring scale, a meter stick, a fulcrum, a set of pulleys, and objects of known weight. To reduce friction, put weights in a "car" that has free-moving wheels for inclined plane experiments. Remember that the weight of the "car" must be included in the load-weight. Each topic in this project can be expanded by adding experiments or combining them with other course material from your own teaching units.
- If you do not understand the concept of using an inclined plane, lever, or pulley to reduce the amount of force needed to perform a certain amount of work, study the handouts and the teacher's edition of your science textbook as a preparation for explaining these things to students.

TERMS AND SYMBOLS REFERENCE SHEET

In many fields of science, special terms and symbols are used to explain concepts and simplify problem solving. This handout provides the basic terms and symbols that you should know to complete the physics lessons. Applications of these terms and symbols are explained more fully in project handouts.

Physics Terms

Inclined plane: a ramp used to reduce the amount of force required to lift a load. It is a simple machine.

Lever: a bar which turns on a fixed support (fulcrum) and is used to reduce the amount of force required to lift a load. It is a simple machine.

Pulley: a wheel with a grooved rim in which a rope can run and so change the direction of a pull. It is a simple machine and is used to raise weights.

Work: transference of energy from one body or system to another, causing motion. Work done is equal to the force exerted times the distance a load is moved.

Load: the object being moved; it is defined by its weight.

Force: the energy needed to move a load.

Fulcrum: support on which a lever turns or rests in moving or lifting a load.

Friction: resistance to motion of surfaces that touch.

Mass: measure of the quantity of matter an object contains. On earth the mass of an object equals its weight.

Physics Symbols

F = force
W = work
D = distance
lb = pounds
d_L = load distance; for a lever, the distance from the fulcrum to the load
d_F = force distance: for a lever, the distance from the fulcrum to the point at which a force is applied

© 1987 by The Center for Applied Research in Education, Inc.

Name _____ Date _____

INCLINED PLANES
Information Sheet

Since the beginning of recorded history, and even before, human beings have searched for ways to make their lives easier, to reduce the amount of *energy* they had to use. One of the earliest machines ever invented was the inclined plane. People discovered that objects too heavy to lift could be slid up a flat, inclined surface. As the example below illustrates, it requires less *force* to move a load on an inclined plane than it does to lift it into a truck.

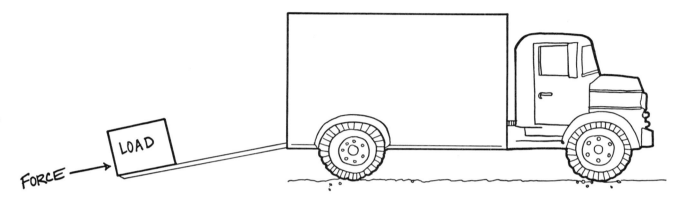

Whether the load is moved up the inclined plane or directly into the truck, the amount of *work* done is exactly the same. The inclined plane, however, reduces the amount of force needed to move the load. The work done is equal to the force exerted times the distance the load is moved.

WORK = FORCE × DISTANCE   *(NOTE:* When you lift something without the aid of an
   W   =   F   ×   D   inclined plane, the *weight* of the object equals the force needed
                         to lift it.)

Example:

The load weighs 200 pounds. The bed of the truck is 4 feet off the ground. Without an inclined plane the load (200 pounds) must be lifted 4 feet. To find how much work is done, follow these steps:

1. Known equation: W = F × D
2. Substitute: W = 200 lb × 4 ft
3. Multiply: W = 800 ft-lb (*NOTE:* The unit for work is "foot-pounds.")

If the load is pushed up an inclined plane that is 10 feet long, the work done is the same. To find the amount of force needed, follow these steps:

1. Known equation: W = F × D
2. Substitute: 800 ft-lb = F × 10 ft
3. Related equation: F = 800 ft-lb ÷ 10 ft
4. Divide: F = 80 lb

The force required to move a 200-pound load up a 10-foot inclined plane is *80 pounds,* while it took 200 pounds of force to lift it 4 feet.

    NOTE: You must calculate the work (W) required to lift a load from the ground to a certain height before you can make calculations for moving it up an inclined plane as shown above.

INCLINED PLANES
Problem Sheet

1. You are trying to get 120 pounds of books up to a 7-foot-high shelf. You have an inclined plane that is 21 feet long. How much force is required to get the books up to the shelf? Make a drawing and show all of your calculations.

2. You are walking up a hill that is 100 feet high. When you get to the top you will have walked 200 feet from the place where you started (at the base of the hill). How much force did you have to exert to walk up the hill? (Use your own weight.) Make a drawing and show all of your calculations.

3. You are sliding down a hill that is 225 feet long. The bottom of the hill is 75 feet lower than the top. How much gravitational force is used to get you down the hill? (Use your own weight.) Make a drawing and show all of your calculations.

© 1987 by The Center for Applied Research in Education, Inc.

Name _____ Date _____

LEVERS
Information Sheet

The lever is another type of simple machine that was discovered thousands of years ago. It was found that a long limb placed over a rock and wedged beneath a heavy object could perform work if people pushed down on the end that extended into the air. This invention allowed things to be lifted, pried, dislodged, and uprooted. Below is an explanation of how levers reduce the amount of force needed to lift a load. The ancients could not have explained the mathematics or the physics involved, but they could have told you that a lever *works*.

I. Levers can change the direction of force or move a load in the same direction as the force.

Example 1:

When the fulcrum is between the load and the force, the lever changes the direction of force.

Example 2:

When the load is between the fulcrum and the force, the lever does not change the direction of force.

- -

II. Levers reduce the amount of force needed to lift a load. The equation for calculating force applies to either example below, regardless of where the fulcrum is placed. The sample problem shows how the equation is used.

Example 3:

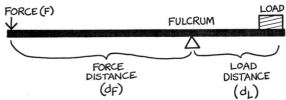

Equation:

$$\text{Force} = \frac{\text{load distance}}{\text{force distance}} \times \text{load}$$

$$F = \frac{d_L}{d_F} \times \text{load}$$

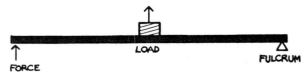

> **Sample Problem:**
>
> $d_L = 35$
> $d_F = 77$
> Load = 420 lb
>
> $F = \dfrac{d_L}{d_F} \times \text{Load}$
>
> $F = \dfrac{35}{77} \times 420\ \text{lb}$
>
> $F = \dfrac{35 \times 420\ \text{lb}}{77}$
>
> $F = \dfrac{14{,}700\ \text{lb}}{77}$
>
> $F = 190.9\ \text{lb}$

Example 4:

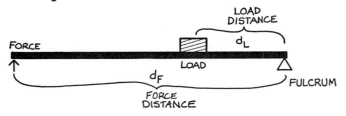

Name _____ Date _____

LEVERS
Problem Sheet

1. How much force is necessary to lift the 120-lb load?

FORCE

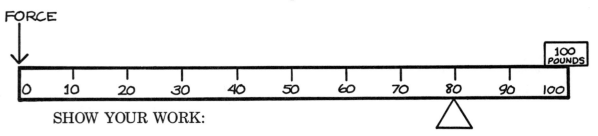

$F = \dfrac{d_L}{d_F} \times$ load (equation for force)

F =

F =

F =

List what you know:

d_L =
d_F =
load =

2. How much force will it take to lift 100 pounds?

FORCE

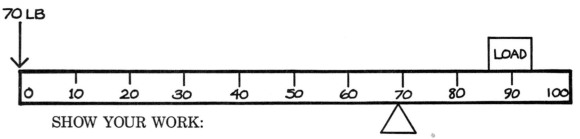

SHOW YOUR WORK:

List what you know:

d_L =
d_F =
load =

3. How heavy a load will 70 pounds of force lift?

70 LB

SHOW YOUR WORK:

List what you know:

d_L =
d_F =
F =

Name _____ Date _____

PULLEYS
Information Sheet

Pulleys were undoubtedly invented sometime after the wheel was first thought of. They were probably originally used simply to change the direction of a force. A pulley mounted over a well allows a bucket of water to be lifted *up* by pulling *down* on a rope. Eventually some inventive mind realized that by using pulleys in combination, the amount of force needed to lift an object could be reduced. The information below shows how to calculate the amount of force needed to lift a load, based on how many pulleys are used and how they are arranged.

I. Pulleys can multiply force (or reduce the amount of force needed to move a load).

II. Pulleys can change the direction of a force.

III. The equation for calculating the amount of force needed to move a load with a set of pulleys is

F = load weight ÷ number of support lines attached to the load

or

$$F = \frac{\text{load weight}}{\text{number of support lines}}$$

IV. If the direction of the force is *opposite* the direction the load moves, the line on which the force is exerted is *not* counted as a support line (see example 1). If the direction of the force is the *same* as the direction the load moves, the line on which the force is exerted *is* counted as a support line (see example 2).

V. To find how far the load will move, divide the distance the rope is pulled by the number of support lines.

Example 1:
Changes direction of force

Example 2:
Does not change direction of force

Example 3:
Changes direction of force

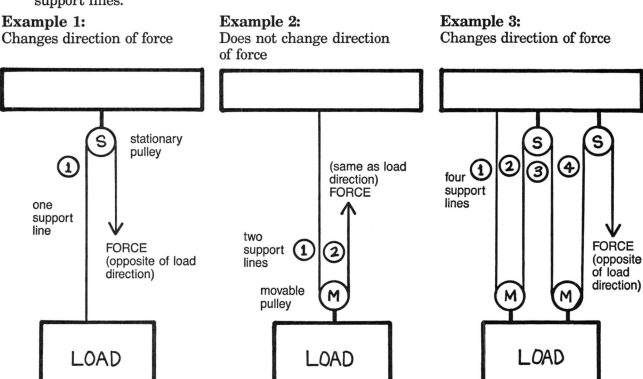

If the load is 50 lb:

$$F = \frac{50 \text{ lb}}{1} = 50 \text{ lb}$$

If the load is 50 lb:

$$F = \frac{50 \text{ lb}}{2} = 25 \text{ lb}$$

If the load is 50 lb:

$$F = \frac{50 \text{ lb}}{4} = 12.5 \text{ lb}$$

11

PULLEYS
Information Sheet (continued)

Answering the following questions will help you better understand each of the three pulley setups described on this handout.

Example 1 questions:

1. If the load is 150 pounds, the force needed to lift it is _____.

2. If the load is _____ then the force needed to lift it is 70 pounds.

3. If you pull the rope 4 feet, the load will move _____ feet.

Example 2 questions:

4. If the load is 150 pounds, the force needed to lift it is _____.

5. If the load is _____ then the force needed to lift it is 70 pounds.

6. If you pull the rope 4 feet, the load will move _____ feet.

Example 3 questions:

7. If the load is 150 pounds, the force needed to lift it is _____.

8. If the load is _____ then the force needed to lift it is 70 pounds.

9. If you pull the rope 4 feet, the load will move _____ feet.

Name _____ Date _____

PULLEYS
Problem Sheet

I.   Mark each *stationary* pulley below by putting an "S" next to it.

II. Mark each *moving* pulley below by putting an "M" next to it.

III. Draw arrows to show the direction each load will move and the direction in which the force on each load is applied.

IV. Solve pulley problems 1–6 below; record your calculations on a separate piece of paper.

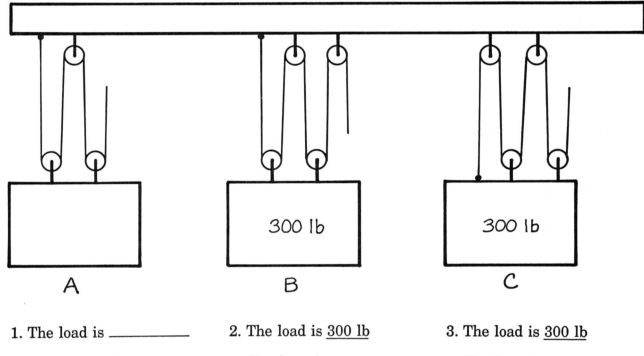

1. The load is _____

   The force is <u>45 lb</u>

2. The load is <u>300 lb</u>

   The force is _____

3. The load is <u>300 lb</u>

   The force is _____

4. Which pulley system above (B or C) requires the least amount of force?_____

5. You have 6 pulleys. On a separate piece of paper sketch the best arrangement for these pulleys.

6. You weigh 80 pounds and you can lift 80 pounds. How large a load can you move with your system of 6 pulleys?_____

Show all of your math work on the paper with your drawing. *Label everything,* showing clearly the load, stationary pulleys, movable pulleys, and direction of the force.

PHYSICS PROJECTS
Teacher's Answer Key

A. INCLINED PLANES Problem Sheet

1. Known equation: $W = F \times D$ (work needed to get 120 lb to the shelf)
    Substitute: $W = 120$ lb $\times 7$ ft (height to lift books)
    Multiply: $W = 840$ ft-lb

    The load is pushed up a 21-foot inclined plane; what force is required?

    Known equation: $F = W \div D$ (*related* to $W = F \times D$)
    Substitute: $F = 840$ ft-lb $\div 21$ ft
    Divide: $F = 40$ lb

2. This is *only an example,* using a student who weighs 150 pounds. Each student is to use his or her own weight to calculate an answer for this problem.

    Known equation: $W = F \times D$ (work needed to get 150 lb to the top of the hill)
    Substitute: $W = 150$ lb $\times 100$ ft (height of the hill)
    Multiply: $W = 15,000$ ft-lb

    The student walks up a 200-foot inclined plane; what is the force required?

    Known equation: $F = W \div D$ (*related* to $W = F \times D$)
    Substitute: $F = 15,000$ ft-lb $\div 200$ ft
    Divide: $F = 75$ lb

3. This is *only an example,* using a student who weighs 150 pounds. Each student is to use his or her own weight to calculate an answer for this problem.

    Known equation: $W = F \times D$ (work to get 150 lb to the bottom of the hill)
    Substitute: $W = 150$ lb $\times 75$ ft (vertical distance to the bottom)
    Multiply: $W = 11,250$ ft-lb

    The student slides down a 225-foot slide or sled run; what is the force required?

    Known equation: $F = W \div D$ (*related* to $W = F \times D$)
    Substitute: $F = 11,250$ ft-lb $\div 225$ ft
    Divide: $F = 50$ lb

B. LEVERS Problem Sheet

1. Known equation: $F = \dfrac{d_L \times \text{load}}{d_F}$

    Substitute: $F = \dfrac{30 \times 120 \text{ lb}}{60}$

    Multiply: $F = \dfrac{3,600 \text{ lb}}{60}$

    Divide: $F = 60$ lb

2. Known equation: $F = \dfrac{d_L \times load}{d_F}$

   Substitute: $F = \dfrac{20 \times 100 \text{ lb}}{80}$

   Multiply: $F = \dfrac{2{,}000 \text{ lb}}{80}$

   Divide: $F = 25 \text{ lb}$

3. Known equation: $F = \dfrac{d_L \times load}{d_F}$

   Related equation: $F \times d_F = d_L \times load$

   $\dfrac{F \times d_F}{d_L} = load$

   Substitute: $Load = \dfrac{70 \text{ lb} \times 70}{20}$

   Multiply: $Load = \dfrac{4{,}900 \text{ lb}}{20}$

   Divide: $Load = 245 \text{ lbs}$

C. PULLEYS Information Sheet

Example 1:
1. $F = 150 \text{ lb}$
2. $Load = 70 \text{ lb}$
3. 4 ft

Example 2:
4. $F = 75 \text{ lb}$
5. $Load = 140 \text{ lb}$
6. 2 ft

Example 3:
7. $F = 37.5 \text{ lb}$
8. $Load = 280 \text{ lb}$
9. 1 ft

D. PULLEYS Problem Sheet

1. Known equation: $F = \dfrac{load}{number \ of \ support \ lines}$

   Related equation: $Load = F \times number \ of \ support \ lines$
   Substitute: $Load = 45 \text{ lb} \times 4$
   Multiply: $Load = 180 \text{ lb}$

2. Known equation: $F = \dfrac{load}{number \ of \ support \ lines}$

   Substitute: $F = \dfrac{300 \text{ lb}}{4}$

   Divide: $F = 75 \text{ lb}$

3. Known equation: $F = \dfrac{\text{load}}{\text{number of support lines}}$

    Substitute: $F = \dfrac{300 \text{ lb}}{5}$

    Divide: $F = 60 \text{ lb}$

4. C

5.

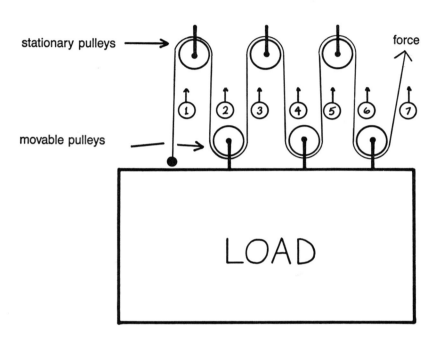

6   Known equation: $F = \dfrac{\text{load}}{\text{number of support lines}}$

    Related equation: Load $= F \times$ number of support lines
    Substitute: Load $= 80 \text{ lb} \times 7$
    Multiply: Load $= 560 \text{ lb}$

ENERGY AND ELECTRICITY

Teacher Preview

Length of Project: 3 hours
Level of Independence: Basic
Goals:

1. To introduce students to a study of electricity.
2. To show students how much energy is used by various electrical appliances.
3. To incorporate mathematics into a science project.

During This Project Students Will:

1. Gain an understanding of the terms "volt," "amp," "watt," "watt-hour," and "kilowatt-hour."
2. Collect data about electricity consumption.
3. Work with equations.
4. Report findings to the class.

Skills:

Listening	Collecting data
Observing	Persistence
Neatness	Personal motivation
Organizing	Sense of "quality"
Identifying problems	Basic mathematics skills
Concentration	Individualized study habits

Handouts Provided:

- "Student Assignment Sheet"
- "Data Sheet"

PROJECT CALENDAR:

HOUR 1: _____	HOUR 2: _____	HOUR 3: _____
Introduction to the terms volt, amp, watt, watt-hour, and kilo-watt-hour. Sample problems are solved, using the equation $W = V \times A$.	Discussion about finding, recording, and calculating information from electrical appliances and tools (amps, volts, and watts). Students receive assignment and data sheets.	Discussion of data brought in by students.
PREPARATION REQUIRED	HANDOUTS PROVIDED	STUDENTS TURN IN WORK
HOUR 4: _____	**HOUR 5:** _____	**HOUR 6:** _____
HOUR 7: _____	**HOUR 8:** _____	**HOUR 9:** _____

Lesson Plans and Notes

HOUR 1: Explain the terms "volt," "amp," "watt," "watt-hour," and "kilowatt-hour." This is not a high school physics class, so the technical aspects of electricity need not be covered; students are simply told that electricity has strength, or power, that is measured in *amps,* and that it is "pushed" into their homes by a force that is measured in *volts*. The amount of work that electricity does is measured in *watts,* and the amount of work done in an hour is measured in *watt-hours*. One thousand watt-hours equals one *kilowatt-hour*. Consumers are charged for their electricity by the kilowatt-hour.

Also, introduce students to the equation $W = V \times A$ (watts = volts × amps), and work some problem examples on the board.

Notes:

- It is assumed that students participating in this project have a basic understanding of equations and related sentences. In other words, they should know that

if $\quad W = V \times A$

then $\quad V = W \div A$

and $\quad A = W \div V$

This is important because they should be able to find data for any two variables and calculate an answer for the third (for a particular electrical appliance or tool). An additional hour should be provided if students need work on solving equations.

- Come to the first class hour prepared with some data from appliances in your own home, or in the school. Use this information to create problems that can be solved with the equation $W = V \times A$.

HOUR 2: Give students the two handouts provided with this project and discuss proper ways of finding information about appliances, and recording it on the data sheet. Show students how to calculate watt-hours per week, watt-hours per month, watt-hours per year, and kilowatt-hours. Their assignment is to take the data sheets home and complete the required research for at least six appliances or tools. These sheets must be brought to class for Hour 3. They will also be required to turn in a math worksheet showing all of their calculations for this project.

HOUR 3: Students bring their homework from Hour 2 to class and use their research findings to discuss the energy consumption of various appliances. A discussion about energy conservation and production would be an obvious addition to this hour's lesson. Ask students how they would react if a situation arose in which electricity had to be rationed. Which appliances or tools could they live without and which would be essential? What electrical machines, other than household appliances, are crucial to the quality of our modern life? At the end of the hour, students turn in their homework.

Notes:

- Students should be required to find out how much one kilowatt hour of electricity costs from the local utility company. They can obtain this information from a family electricity bill or from a phone call to the power company. A figure for cost per kilowatt-hour is necessary so students can calculate how expensive certain appliances or tools are to operate per day, week, month, or year.

- An additional hour may be needed if time is to be spent in class calculating how much it costs to operate each student's list of appliances and tools. Three hours is sufficient if only a few examples are to be discussed in class, but, obviously, the project can be expanded to include increased emphasis on the data that students bring in. For example, charts can be made to compare the amount of electricity consumed by certain appliances, or a composite "household" can be developed to illustrate electricity consumption throughout a home. There are many directions this project can go once basic data collection has been completed.

ENERGY AND ELECTRICITY
Student Assignment Sheet

Electricity is a basic source of energy available to most people in America. It is so basic, in fact, that we take it for granted and just assume it will always be there. Yet *somebody* has to provide electricity, and it has a cost. If the bills aren't paid, lights go dark, stereos stop playing, refrigerators stop cooling, televisions lose sight and sound, computers shut down—things aren't the same without electricity. For this project you will examine a number of appliances, machines, and tools in your home to see how much electricity they use and how much it costs to operate them. This activity should provide some insight into the importance of electricity in your life (just add up the number of electricity-consuming items in your house), and into the mathematics involved in making calculations about electricity consumption and cost. Electricity is an important area in the study of *physics,* so you are also being introduced to that branch of science.

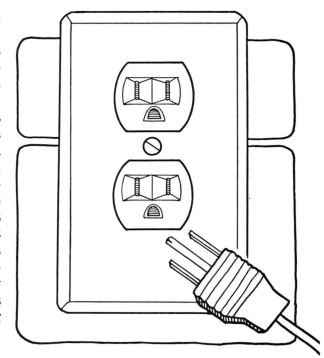

Here is your assignment.

I. On the energy and electricity "Data Sheet":
  A. List at least six items in your home that run on electricity.
  B. Under the appropriate heading, record pertinent information you find at home about each item. You are looking for volts, amps, or watts. Find two of these, and then use the equation "watts = volts × amps" from the data sheet to calculate the third. Write "given" in the box with each piece of information that you find written on an appliance. An appliance example is done for you on the data sheet handout. The voltage and wattage information was found written on the cassette tape deck, so the word "given" is recorded next to these entries. Then the equation A = W/V was used to calculate amps for this item.
  C. Estimate and record how many hours each item is used in a typical day.
  D. Also calculate and record for each item:
    1. Watt-hours per day
    2. Watt-hours per year
    3. Kilowatt-hours per year
    4. Cost per year for the electricity used (You will need to find out how much one kilowatt of electricity costs.)
II. On a separate "math worksheet" show all the calculations for your data sheet entries in an orderly way. The answers should be labeled with units.
III. Be prepared to present your work to the class if called upon.
IV. Turn in your completed energy and electricity "Data Sheet" and the math worksheet which shows all of your calculations.

ENERGY AND ELECTRICITY DATA SHEET

VOLTS: The "force" that pushes electricity through a wire.

AMPS: The "strength" of an electric current; the amount of current being pushed through a wire.

WATTS: Unit of electric power or energy.

WATT-HOUR: The amount of work one watt of electricity can do in one hour.

COST PER KILOWATT-HOUR: This is how you are billed for the electricity you use at home. In your area cost per kilowatt hour = ————.

KILOWATT-HOURS: $\dfrac{\text{watt-hours}}{1000}$

1. WATTS = VOLTS × AMPS
   W = V × A

2. WATT HOURS = W × T (timed used)

3. Most things that plug into a wall socket use 110 volts.

4. Some large appliances and motors use 220 volts.

APPLIANCE OR TOOL	VOLTS $V = \frac{W}{A}$	AMPS $A = \frac{W}{V}$	WATTS $W = V \times A$	HOURS OF USE/DAY ESTIMATED	WATT-HOURS/ DAY	WATT-HOURS/ YEAR	KILOWATT-HOURS/YEAR	COST/YEAR
CASSETTE TAPE DECK	GIVEN: 120	CALCULATED: .075	GIVEN: 9	3	27	9855	9.855	$0.70
1.								
2.								
3.								
4.								
5.								
6.								
TOTALS:								

SPEED OF TOBOGGANS

Teacher Preview

General Explanation:

This project is based upon a basic principle of physics: that the acceleration of gravity is the same for all freely falling bodies, regardless of their mass. Applied to this project, the principle says that, under *ideal* conditions, no matter how much weight is on a toboggan, it will always require the same amount of time to cover a given distance on a particular slope. Students generally hypothesize that adding more weight means it will go faster. Field data will be less than perfect so lessons are based on discussing factors that prevented the field experiment from being ideal and on conducting a more ideal experiment indoors to see what the results will be.

Length of Project: 4 hours

Level of Independence: Basic

Goals:

1. To conduct a science experiment with students outdoors.
2. To emphasize independent thinking.
3. To provide an opportunity for students to see how certain science problems are solved.
4. To incorporate mathematics into a science project.

During This Project Students Will:

1. Discover how the terms "distance," "time," and "speed" are related to "weight" (or mass).
2. Hypothesize the outcome of an experiment.
3. Record data (distance, time, and weight) about the speed of their toboggans (or wagons).
4. Calculate average speeds.
5. Graph the results of their field data.
6. Work in small groups to complete this project (five students per group).
7. Analyze experimental data and conditions.

Skills:

Collecting data	Following project outlines
Listening	Individualized study habits

Observing   Persistence
Neatness   Taking care of materials
Group planning   Personal motivation
Identifying problems   Self-awareness
Working with others   Sense of "quality"
Accepting responsibility   Basic mathematics skills
Concentration   Drawing/sketching/graphing
Controlling behavior   Organizing

Handouts Provided:

- "Student Assignment Sheet"
- "Speed of Toboggans Worksheet"

PROJECT CALENDAR:

HOUR 1:	HOUR 2:	HOUR 3:
Introduction to the assignment sheet and a step-by-step explanation of the experiment.	The experiment is conducted outdoors on a hillside, with students recording data on worksheets. Worksheets are taken home to be finished.	Discussion of the completed worksheets.
PREPARATION REQUIRED HANDOUT PROVIDED	HANDOUT PROVIDED NEED SPECIAL MATERIALS	STUDENTS TURN IN WORK

HOUR 4:	HOUR 5:	HOUR 6:
A classroom experiment is conducted under the best possible conditions to see what the results of the toboggan experiment should ideally be.		
NEED SPECIAL MATERIALS RETURN STUDENT WORK PREPARATION REQUIRED		

HOUR 7:	HOUR 8:	HOUR 9:

Lesson Plans and Notes

HOUR 1: Give students the assignment sheet, which includes a worksheet and graph. Divide the class into small groups and explain the project step-by-step, with special attention to accuracy in measuring and recording data. Carefully describe the procedure for conducting the outdoor experiment and explain safety precautions. Each student records his or her hypothesis on the assignment sheet.

Note:

- If there is no record of the students' weight available, a scale should be provided and the last part of the hour spent weighing students to help emphasize accurate data collection. Each student should record his or her own weight. It is not a bad idea to have them wear the boots and coats they will wear during Hour 2. Perhaps a parent or student aide can assist with this, or maybe students can simply be allowed to leave class one at a time to weigh themselves. If it is not possible to provide a scale, estimated weights will work fine.

HOUR 2: Conduct the speed of toboggans experiment outdoors. Each group should follow this procedure:

1. Establish a distinct starting line at the top of the hill and mark a finish line at the bottom of the hill. Measure the distance (d) between these two lines and record it on the worksheet.

2. One group member (at the top of the hill) waves a flag just as the front of the toboggan crosses the starting line.

3. Another group member (at the bottom of the hill) starts timing with a stopwatch at the instant the flag waves, and stops the watch when the front of the toboggan crosses the finish line. This person has a worksheet on a clipboard to record the time for each run.

4. The remaining three group members are toboggan riders.

5. At the end of this hour's project, each group member copies all of the data from the group worksheet onto a separate worksheet that can be taken home. Completing the graph by calculating average times and speeds is a homework assignment. Each student brings his or her own completed worksheet to class the next hour.

6. Proper safety precautions must be taken for this activity: select a safe hill, and use toboggans that are in good shape. Instruct students about safety procedures (given below). Devise an organized way for students to get back to the top of the hill without being in the path of students coming down.

SAFETY PROCEDURES

1. Arms and legs must be *inside* the toboggan.
2. The toboggan run should be a gradual grade so that there is no danger of capsizing the toboggan.
3. Toboggans should be large enough to accommodate three students.
4. Students must wear mittens and head coverings such as stocking caps or jacket hoods for protection.
5. Toboggans are started with a nudge; "speed pushes" are not allowed at any time.
6. Students follow a predetermined path back to the top of the hill. This path begins at the end of the toboggan run, leads *away* from the other runs, and goes up the hill at a place that is separated from the toboggan runs.
7. Students must remove pencils and other sharp objects from their pockets before they get on toboggans.
8. An area for timers and flag wavers needs to be determined for each run. This area should be off to the side of the run so these people don't interfere with the toboggans, or vice versa.
9. Students should not chew gum.

Notes.

- Students graph the results of their experiment according to the amount of weight on the toboggan for each trial run. The handout suggests that the toboggan be timed going down the hill with no one on it, with one person, then two, and then three. A lot more weight possibilities than this exist, however. For example, if three students weigh

  A. 86 pounds
  B. 97 pounds
  C. 115 pounds

  then three trials, one for each person, could be run, followed by various combinations of increasing weight. You could do trials with A and B (183 pounds), A and C (201 pounds), B and C (212 pounds), and A, B, and C (298 pounds). This would give seven sets of trial runs and thus a more representative graph.

- If the results are not perfect (which they won't be), require students to list the factors that worked against perfect conditions: friction, wind, too much snow, too little snow, sliding off the measured run, sled turning sideways, and so forth.

- You will need this equipment:

  Stopwatches (1 for each group)
  Tape measure (30-meter or 100-foot, if possible)
  Flags (1 per group)
  Clipboards (1 per group)
  Toboggans or sleds (1 per group)

- The graph that is provided on the worksheet has "Weight in Pounds" across the bottom axis (ranging from 0 to 480 pounds). The vertical axis is not filled in with numbers for average speed, since varying conditions will yield a range of possible speeds. Students should fill in this part of the graph *after* they have conducted the experiment, putting their slowest speed at the bottom, their fastest speed at the top, and the difference between the two divided into equal increments. In a *perfect* experiment, each trial run would take exactly the same amount of time, and the speed would be the same for every weight. In this case, the graph is a straight horizontal line that should be placed in the middle of the graph. In actual practice, however, the results will be far from perfect and the graph will show a range of speeds.

- The vertical axis of the graph ("Average Speed") has no units given to it. Student should supply this information. It may be feet per second, meters per second, or yards per second.

- Obviously, "Speed of Toboggans" was created for a geographic area that has a snowy winter season; wagons, soapbox derby cars, or some other wheeled vehicles must be used in warm climates or during warm seasons. If this project is to be done with wheeled vehicles, think the process through carefully to ensure a successful experiment. Safety precautions are especially critical when using wheeled vehicles!

HOUR 3: Students bring their completed worksheets to class and spend the hour discussing the results. Worksheets are handed in at the end of the hour.

Note:

- Don't tell students what the ideal results of the toboggan experiment should be. Suggest that they conduct a related indoor experiment (Hour 4), under more perfect conditions. Ask students to write another hypothesis before coming to Hour 4.

HOUR 4: Conduct a controlled version of the experiment in the classroom this hour, using a car with "frictionless" wheels, and a set of accurate weights. Students time three trial runs of the car without any weights (after the length of the ramp is measured), and the times are recorded on the chalkboard. Various amounts of weight are then placed on the car and trial runs conducted; three trials for each amount of weight. Students calculate the average time (t) and average speed (s) for each amount of weight used. This information is then graphed, either in large scale on the board or by each student individually on graph paper. Appropriate scales for weight and average speed must be made.

Decide how to make this graph before class. Under ideal conditions the graph will be a straight horizontal line. At the end of this hour, return the completed worksheets handed in during Hour 3.

Notes:

- You will need this equipment:
    Stopwatch
    Tape measure
    Flag
    Ramp
    Small car
    Weights (for the car)
- A suggestion for the classroom experiment is to use an electric train car (a flat car or a coal car) and several lengths of straight track fastened to an eight- or ten-foot pine board.
- With some modification, the "worksheet" can also be used for the in-class experiment.

SPEED OF TOBOGGANS
Student Assignment Sheet

Experiments are a scientist's way of investigating how nature works: a way of learning from what one observes. It is important for anyone interested in science to understand how to carefully set up an experiment and accurately record everything that happens. "Speed of Toboggans" is an experiment, and you are the scientist. As you conduct the experiment, keep your mind focused on the *purpose* of such an undertaking: you are learning how to state a hypothesis, set up an investigation, record data, and draw conclusions from the results of your work.

This project will be conducted in groups, with five students per group. Your group will investigate the relationship between mass (which means the amount of "weight" you put on the toboggan) and how fast the toboggan travels down the hill. As weight is added will the toboggan require more, less, or the same amount of time to travel down the hill? Before you begin, write your hypothesis (your guess) as to when the toboggan will travel fastest: when the *least* amount of weight is on the toboggan, when the *most* weight is on it, or will it always require the same amount of time, regardless of weight?

HYPOTHESIS: _____

_____

When conducting an investigation all factors except one should be constant; there should be only one variable. The constant factors, those that will remain the same, will be the slope of the hill, the condition of the hill, the length of the run, and the toboggan itself. Try to keep these control factors from varying as much as possible by staying in the same area of the hill for each trial run. Your *variable* factor will be the weight factor—of you and your friends in the group. The toboggan "run" will be timed from the starting point until the end of the measured run. Have one person in your group be responsible for recording data.

PROCEDURE:

I. Measure the length of the run in meters, feet, or yards. Record this distance on your worksheet.

II. Send your toboggan down the hill without anyone on it, and record its time. Let gravity pull it down the hill by nudging it just enough to get it moving at the starting line. Repeat this procedure two more times.

III. Send the toboggan down the hill with one person on it and record its time. Once again, just nudge it to get it started. Repeat this procedure two more times with the same person on the toboggan. Be sure to record the weight of the person who rode the toboggan.

IV. Send the toboggan down the hill with two people in it, and record its time. Repeat this procedure two more times with the same two people. Record the *combined* weight of these two people.

V. Send the toboggan down the hill with three people on it, and record its time. Repeat this procedure two more times. Record the combined weight of all three people.

VI. You may do more trial runs with different amounts of weight if you wish. Record all data from these extra runs on the worksheet.

VII. Fill out the "Speed of Toboggans Worksheet." To calculate the average speed of your toboggan for the various amounts of weight that were used, use this equation: $s = d \div t$.

Notes:

s = speed (the *average speed* of 3 runs)
d = distance (recorded earlier)
t = time (t = *average time* for each set of three trial runs from the "Speed of Toboggans Worksheet")

Show all of your calculations on a separate sheet of paper. BE NEAT!

VIII. Bring your work to class where the experiment will be discussed.

Name _____ Date _____

SPEED OF TOBOGGANS WORKSHEET

Members in your group: (Include yourself)

1. Flag person _____

2. Timer _____

3. Tobogganer _____ Weight: _____ lb

4. Tobogganer _____ Weight: _____ lb

5. Tobogganer _____ Weight: _____ lb

DATA CHART

Distance of run (d) = _____

Record the *time* (in seconds) and *weight* (in pounds) for each trial on the following chart, then calculate the average time and average speed for each set of three trial runs.

		AMOUNT OF WEIGHT ON THE TOBOGGAN							
		toboggan only	___ lb	___ lb	___ lb	___ lb	___ lb	___ lb	___ lb
time in seconds	TRIAL 1								
	TRIAL 2								
	TRIAL 3								
	average time (t)								
	average speed (s)								

DATA GRAPH

Graph the results of your investigation with a *line graph* below.

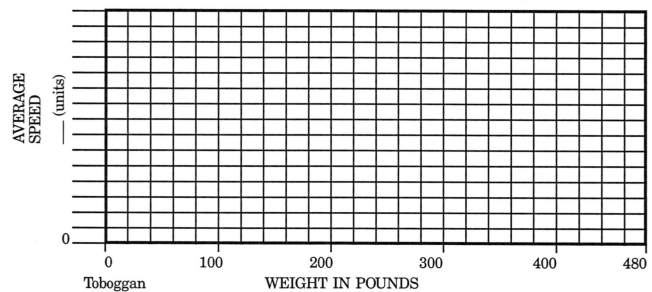

© 1987 by The Center for Applied Research in Education, Inc.

CHEMISTRY CONCEPTS
FOR THE TEACHER

This is a brief review of concepts covered in Project S-4. This learning project is not designed to teach high school chemistry. Motivated upper elementary and junior high school students can, however, be taught a great deal about chemistry before they reach high school if you follow a logical progression, move through the material slowly, review information often, and explain concepts clearly. In order to teach chemistry you must understand the concepts yourself; you need not be a chemistry major or a chemist, but it is important to take some time to learn (or relearn) the details of what is to be taught. A concise explanation of the areas of chemistry that are covered in this project is provided here. Also, study the lesson plans, notes, and student handouts. A high school chemistry textbook would be a good reference if you have questions or do not understand certain concepts. Or, talk to a high school or college chemistry instructor to clarify the areas you are unsure of.

The areas covered in this 16-hour chemistry project are

1. Atomic structure
2. Periodic table of the elements
3. Ionic bonding
4. Acid-base reactions

Each of these areas is explained in this introductory material. Remember that, although this information is accurate, it is presented in simplified form. Students should clearly understand that the purpose of this project is to *prepare* them for their first chemistry course in high school. They should also understand that they are not learning all there is to know about atomic structure, ionic bonding, and so forth, but they are laying a solid foundation from which they can learn more difficult concepts.

1. ATOMIC STRUCTURE

An atom is made of three primary components: protons, neutrons, and electrons. The protons and neutrons are in the nucleus, and electrons are located in orbitals, or shells, around the nucleus. Protons are positively charged ($+$) particles and electrons are negatively charged ($-$) particles; protons, however, are nearly 2,000 times larger than electrons. Since protons and neutrons are the same size they determine the weight of an atom. For this project the electrons are said to be in their "principal" shells; the word orbital is not used. The first principal shell can contain up to two electrons, the second principal shell can contain up to eight electrons and the third principal shell can contain up to eighteen electrons. From now on the principal electron shells will be referred to simply as "electron shells." The "second electron shell" will refer to the "second principal electron shell."

Students are taught to draw atoms as a series of concentric circles (see "Drawings of Atoms" student handout): the first circle represents the nucleus, and the number of protons and neutrons is recorded in this circle. The second circle is the *first electron shell* which contains the first two electrons in the atom. The third circle is the *second electron shell,* which contains the atom's next eight electrons. The fourth circle is the *third electron shell,* which can contain up to eighteen electrons. However, (and this is *important!*) we are interested only in the *first eight electrons in the third shell.* The eight electrons in the second shell and the first eight electrons in the third shell are called "octets," and this will be important later.

An atom can be drawn if its atomic weight and atomic number are known. The atomic number indicates how many protons the atom has. This number also tells how many electrons the atom has. By rounding the atomic weight (taken from the periodic table of the elements) to the nearest whole number and subtracting the atomic number, the number of neutrons in the atom can be discovered. This is all the information that is necessary to make concentric-circle drawings of atoms, remembering that the first shell can have no more than two electrons and the second and third shells can have no more than eight electrons each.

Example: Aluminum

Atomic number   = 13 (13 protons and 13 electrons)
Atomic weight   = 26.98 or 27 (27 protons + neutrons)
Number of neutrons = atomic weight − atomic number
                  = 27 − 13
                  = 14 neutrons

CONCENTRIC-CIRCLE REPRESENTATION OF ALUMINUM

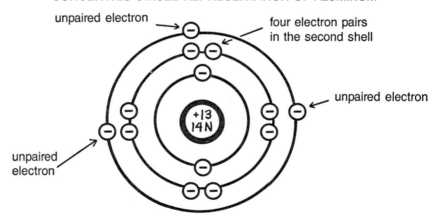

The first four electrons to be placed in either the second or third electron shell are spaced as far apart as possible. The fifth electron "pairs" itself with one of the first four; the sixth electron pairs itself with one of the remaining unpaired electrons; the seventh electron pairs itself with one of the two remaining electrons, and the eighth electron pairs itself with the final unpaired electron. If more electrons are to be added, they must be placed in the next shell because for this project the second and third electron shells can each take eight electrons and no more: an octet. These "electron pairs" are not chemically reactive. Students are told that they can tell how an atom will react with other atoms by observing how many *unpaired* electrons are in its outer shell. (See the "Drawings of Atoms" student handout for additional examples.)

The principal electron shells are actually divided into subshells, which are not discussed in this project, but which are mentioned here for your information. There is one subshell for the first principal electron shell. It can contain two electrons and is called the "1s" subshell. (Essentially the "1s" subshell and the first principal shell are the *same thing*.) The second principal electron shell is divided into two subshells: "2s" and "2p." The "2s" subshell can contain two electrons and the "2p" subshell can contain six electrons. The third principal shell has three subshells: "3s," "3p," and "3d". The "3s" subshell can contain two electrons, the "3p" subshell can contain six electrons and the "3d" subshell can contain ten electrons. The "octet" spoken of earlier refers to the eight electrons in the "s" and "p" subshells for both the second and third principal electron shells.

Although the project does not require this of students, if you want to represent the atomic structure of aluminum by showing how electrons are placed in their subshells, it is done like this (each arrow represents an electron):

ELECTRON CONFIGURATION OF ALUMINUM: Al

Aluminum: Al (atomic number = 13)

3 electrons in the third principal shell

8 electrons in the second principal shell

2 electrons in the first principal shell

Note: Each arrow represents an electron. Paired electrons are illustrated by arrows pointing in opposite directions. This is done because the two electrons are "spinning" in opposite directions.

IMPORTANT: In atoms that have more than eighteen electrons, such as potassium and calcium, the concentric-circle drawings show electrons going into the fourth principal electron shell (because the second and third shells are filled with eight electrons each); these electrons are actually going into the "3d" subshell of the third principal shell. In this case the "3d" subshell is actually the outer shell, even though it is not a principal electron shell. For example, take a look at the following illustrations of the calcium atom:

CONCENTRIC CIRCLE REPRESENTATION OF CALCIUM

Calcium: Ca (atomic number = 20)

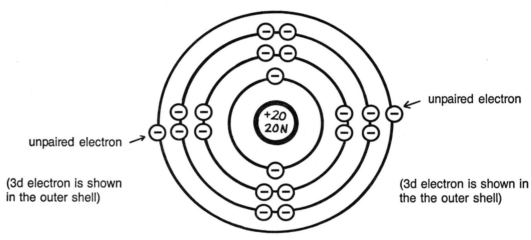

unpaired electron

unpaired electron

(3d electron is shown in the outer shell)

(3d electron is shown in the the outer shell)

ELECTRON CONFIGURATION OF CALCIUM

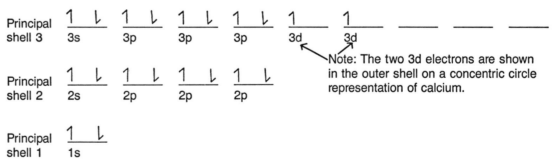

Although you may decide to teach this material in your chemistry class, students need not be introduced to electron subshells to understand the basic concept of atomic reactions and ionic bonding. The important point is this: regardless of how you represent it, calcium has two reactive electrons, which means it has two outer-shell electrons. These electrons determine how calcium behaves chemically.

2. PERIODIC TABLE OF THE ELEMENTS

The periodic table provides important information about pure elements and their atomic structures. The "atomic number" tells how many protons an atom of a given element has. This, in turn, tells how many electrons the atom must have (no. of protons = no. of electrons). The atomic weight tells how many protons and neutrons the atom has *on average*. The number of neutrons can vary, so an average is used to determine the weight of an atom. Since the average is usually not a whole number, atomic weight is expressed as a decimal. To calculate the number of neutrons in an atom, round the atomic weight to the nearest whole number and subtract the atomic number.

The partial table provided with this project as a student handout charts the first twenty elements (through calcium), and a few of the heavy metals to show how large some atoms become. The columns represent valence numbers or, in other words, they tell you how many electrons are in the outer shell of the elements in that group. From left to right, each element has one electron and one proton more than the element that precedes it on the table. All of the elements in group IA have one electron in their outer shells. The elements in group IIA have two outer-shell electrons and so on to the VIIIA elements which have eight outer-shell electrons, or an "octet." The elements in VIIIA are inert because they have no reactive, or unpaired, outer-shell electrons. Their outer shells are filled.

By looking at the table it is possible to tell what reactions will probably take place between elements. There are many possible combinations. Some examples are included here in the form of "Lewis dot formulas," (or "dot configurations"), which represent each element's *outer*-shell electrons.

a. Lithium has one outer electron (IA element): fluorine has seven outer-shell electrons (VIIA element) and needs one more to complete its octet. Therefore:

$$\text{Li}^{\cdot} + \cdot \overset{\cdot\cdot}{\underset{\cdot\cdot}{\text{F}}}\text{:} \rightarrow \text{Li} \overset{\cdot\cdot}{\underset{\cdot\cdot}{\text{:F}}}\text{:}$$

b. Hydrogen has one outer-shell electron (IA); oxygen has six outershell electrons (VIA) and needs two more to complete its octet. Therefore:

$$2\text{H}^{\cdot} + \cdot \overset{\cdot\cdot}{\underset{\cdot\cdot}{\text{O}}}{}^{\cdot} \rightarrow \text{H} \overset{\cdot\cdot}{\underset{\cdot\cdot}{\text{:O}}}\text{:}$$
$$\text{H}$$

c. Magnesium has two outer-shell electrons (IIA); chlorine has seven outer-shell electrons (VIIA) and needs one electron to complete its octet. Therefore:

$$^{\cdot}\text{Mg}^{\cdot} + 2 \overset{\cdot\cdot}{\underset{\cdot\cdot}{\text{Cl}}}{}^{\cdot} \rightarrow \overset{\cdot\cdot}{\underset{\cdot\cdot}{\text{:Cl}}}\text{:Mg} \overset{\cdot\cdot}{\underset{\cdot\cdot}{\text{:Cl}}}\text{:}$$

3. IONIC BONDING

All molecules are formed by one of two types of chemical bonds. Ionic bonding is one type; there is also covalent bonding, which is not covered in this project except to explain the polar water molecule's effect on ionic crystals. Ionic bonding is based upon the loss and gain of electrons, while covalent bonding is based on electron sharing. It is important to understand that many bonds are not purely ionic or covalent, but exhibit characteristics of both. For this project, students will work only with molecules that are considered ionic.

When an atom such as sodium comes in contact with an atom such as chlorine, an ionic reaction takes place: $Na + Cl \rightarrow Na^{+1} + Cl^{-1}$. The chlorine atom, with seven outer shell electrons, takes the single outer shell electron of the sodium atom. This gain of one electron leaves the chlorine with a filled outer shell, but with one more electron than protons. Thus a chloride *ion* is formed with a charge of ⁻1. The sodium atom gives up an electron and becomes a sodium ion with a charge of ⁺1 (one more proton than electrons). These oppositely charged ions form a solid because they attract each other with a strong force that gives the compound *sodium chloride* (NaCl) its stability. Positively charged sodium ions, Na^+, and negatively charged chloride ions, Cl^-, group themselves in a three-dimensional arrangement, creating a crystal of ionic sodium chloride.

A water molecule, on the other hand, is formed by covalent (or electron sharing) bonds between one oxygen atom and two hydrogen atoms. The result of this type of bonding is that the two protons of the hydrogen atoms are concentrated on one side of the molecule, which gives that side a partially positive charge. The outer shell electrons of the oxygen are concentrated on the other side of the molecule which gives that side a partially negative charge. For this reason the water molecule is said to be *polar*, because one side is positive and the other side is negative (see "Polar Water Molecule" student handout). The strength of the water molecule's polarity is great enough to break many ionic crystals. This is why sodium chloride dissolves in water, yielding free ions in solution: the positive ions (Na^{+1}) arrange themselves around the *negative* side of a water molecule and the negative ions (Cl^{-1}) are attracted to the *positive* side of a water molecule.

Another example of this process is illustrated on the drawing of lithium and fluorine. This reaction is twofold: Li + Fl yields an electron transfer, which in turn yields an ionic crystal (lithium fluoride). When the LiFl crystal is dissolved in water, free ions are produced. The free ions, as they are shown, will appear separately only when dissolved in water.

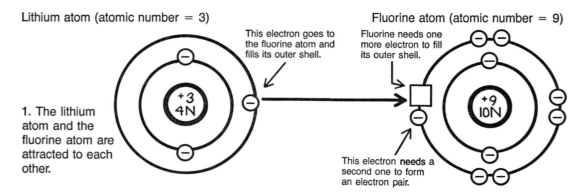

Lithium atom (atomic number = 3)

Fluorine atom (atomic number = 9)

This electron goes to the fluorine atom and fills its outer shell.

Fluorine needs one more electron to fill its outer shell.

1. The lithium atom and the fluorine atom are attracted to each other.

This electron needs a second one to form an electron pair.

The dashed line represents an electron transfer.

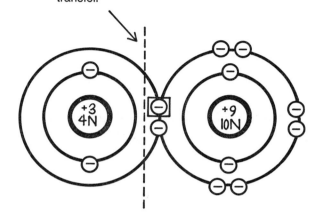

2. Upon coming together, an electron is transferred (from the lithium to the fluorine).

NOTE: After the transfer, electrostatic forces hold the ions together to form a crystal.

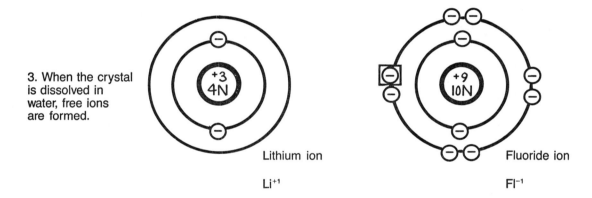

3. When the crystal is dissolved in water, free ions are formed.

Lithium ion

Li^{+1}

Fluoride ion

Fl^{-1}

The lithium and fluoride ions shown in the example form an ionic crystal that, like all ionic crystals, is held together by electrostatic forces between the ions. These forces are not as strong as the forces exerted by the polarity of water molecules and so lithium fluoride will dissolve in water.

4. ACID-BASE REACTIONS

An acid is a substance that yields hydrogen ions (H^+) when dissolved in water. For example, hydrogen chloride, HCl, when dissolved in water, yields: $HCl \xrightarrow{H_2O} H^{+1} + Cl^{-1}$. This is hydrochloric acid.

A base is a substance that yields hydroxide ions (OH^-) when dissolved in water. For example, sodium hydroxide, NaOH, when dissolved in water, yields:

$$NaOH \xrightarrow{H_2O} Na^{+1} + OH^{-1}$$

Acids and bases react with one another to form (1) water, and (2) ionic salts. This is called a neutralization reaction because the acid and the base are both neutralized into water and salt. In solution, the free H^{+1} ions from the acid are powerfully attracted to the free OH^{-1} from the base to form H_2O. The positive ions from the base (ions formed from elements found on the *left* side of the periodic table: IA, IIA and IIIA) and the negative ions of the acid (ions formed from elements found on the *right* side of the periodic table: VA, VIA and VIIA) form a salt in solution (see "Acid-Base Reactions" student handout). If the water is evaporated away, the salt remains.

This project emphasizes simple acid-base equations and nothing more. Since the H^{+1} ion(s) from the acid will combine with the OH^{-1} ion(s) from the base to form water, the only requirement of students is to identify the resulting salt and balance the right side of the equation:

1. $HCL + NaOH \rightarrow H^{+1} + :\overset{..}{\underset{..}{Cl}}:^{-1} + Na^{+1} + :\overset{..}{\underset{..}{O}}:H^{-1} \rightarrow NaCl + H_2O$

2. $H_2S + 2KOH \rightarrow 2H^{+1} + :\overset{..}{\underset{..}{S}}:^{-2} + 2K^{+1} + 2:\overset{..}{\underset{..}{O}}:H^{-1} \rightarrow K_2S + 2H_2O$

3. $2HCl + Ca(OH)_2 \rightarrow 2H^{+1} + 2:\overset{..}{\underset{..}{Cl}}:^{-1} + Ca^{+2} + 2:\overset{..}{\underset{..}{O}}:H^{-1} \rightarrow CaCl_2 + 2H_2O$

The base always yields a *positive* ion of an element from the left side of the periodic table; the acid always yields a *negative* ion of an element from the right side of the periodic table. The salt that is formed is written with the *positive* ion *first: NaCl, K_2S, $CaCl_2$*; or, *sodium* chloride, *potassium* sulfide, *calcium* chloride.

CHEMISTRY PROJECTS

Teacher Preview

Length of Project: 18 hours
Level of Independence: Basic
Projects Offered: Atomic Structure
                    The Periodic Table of the Elements
                    Ionic Bonding
                    Acid-Base Reactions

General Description:

This project is based primarily upon classroom lecture and simple drawings of atomic reactions. Drawings are made on the board (or on transparencies) and students are supplied with a number of illustrative handouts. Note taking is emphasized in teaching students how to interpret the periodic table, draw atoms, predict reactions, and solve acid-base equations. A special section, called "Chemistry Concepts for the Teacher," precedes this project and explains the four areas of chemistry that are covered with students. If you understand this material you should be able to teach this course.

Goals:

1. To prepare students for their first high school chemistry course.
2. To introduce students to a basic science discipline.
3. To emphasize note taking, listening, concentration, and other study skills.
4. To demonstrate that upper elementary and junior high school students are capable of learning about chemistry.

During This Project Students Will:

1. Be introduced to a number of chemistry terms.
2. Learn about the parts, or components, of an atom.
3. Draw concentric-circle representations of atoms and ions.
4. Use the periodic table of the elements to determine the number of protons, neutrons, and electrons in the atoms of any element.
5. Demonstrate the ability to tell how many "outer-shell" electrons an atom has by locating it on the periodic table.
6. Solve problems that involve volume, mass, and density.
7. Study ionic substances and how they are formed.

8. Study the polar water molecule, and how such molecules dissolve ionic substances.

9. Learn a simple definition for an acid and a base.

10. Predict the outcomes of simple acid-base reactions.

11. Solve problems relating to ionic bonding and density.

Skills:

Collecting data	Controlling behavior
Listening	Individualized study habits
Observing	Persistence
Neatness	Personal motivation
Spelling	Sense of "quality"
Organizing	Setting personal goals
Identifying problems	Self-confidence
Concentration	Drawing/sketching/graphing
Basic mathematics skills	Self-awareness

Handouts Provided:

- "Chemistry Lesson 1: Atomic Structure"
- "Chemistry Lesson 2: The Periodic Table"
- "Periodic Table of the Elements (Elements 1 through 20)"
- "Drawings of Atoms: Concentric-Circle Representations"
- "Chemistry Lesson 3: Ionic Bonding"
- "Concentric-Circle Representation of the Water Molecule"
- "Ionization Example 1: Sodium Chloride"
- "Ionization Example 2: Magnesium Oxide"
- "Ionization Example 3: Lithium Oxide"
- "Ionization Example 4: Calcium Chloride"
- "Acid-Base Reaction: Hydrogen Chloride + Sodium Hydroxide"
- "Individualized Classroom Experiment: Electrolysis"
- "Chemistry Final Test" (answers provided)

PROJECT CALENDAR:

HOUR 1: _____	**HOUR 2:** _____	**HOUR 3:** _____
Introduction to chemistry: what it is, what it is used for.	Discussion of three terms: element, atom, molecule.	Discussion of six terms: proton, electron, neutron, nucleus, electron shell, octet. Students are shown how to make concentric-circle drawings of atoms.
PREPARATION REQUIRED	HANDOUT PROVIDED PREPARATION REQUIRED	PREPARATION REQUIRED
HOUR 4: _____	**HOUR 5:** _____	**HOUR 6:** _____
Students make concentric-circle drawings of atoms.	Discussion of the terms atomic number and atomic weight, followed by an examination of the periodic table.	Discussion of the term density. Problems are solved in class.
PREPARATION REQUIRED STUDENTS TURN IN WORK	HANDOUT PROVIDED RETURN STUDENT WORK PREPARATION REQUIRED	HANDOUT PROVIDED STUDENTS TURN IN WORK PREPARATION REQUIRED
HOUR 7: _____	**HOUR 8:** _____	**HOUR 9:** _____
Students are shown how to use information from the periodic table to draw atoms of elements. Drawings are done on the board by students.	At their desks, on their own, students draw concentric-circle representations of several elements, using the periodic table for reference.	Explanation of the polar water molecule and how it dissolves ionic salts.
RETURN STUDENT WORK HANDOUT PROVIDED PREPARATION REQUIRED	PREPARATION REQUIRED STUDENTS TURN IN WORK	RETURN STUDENT WORK HANDOUT PROVIDED

PROJECT CALENDAR:

HOUR 10: _____ Ionization is discussed and illustrations are drawn on the board. **HANDOUTS PROVIDED** **PREPARATION REQUIRED**	**HOUR 11:** _____ Students work on their own, making concentric-circle drawings of atoms and ions. **STUDENTS TURN IN WORK**	**HOUR 12:** _____ Discussion of acids and bases: what each is and the reaction that results from combining an acid and a base. **HANDOUT PROVIDED** **RETURN STUDENT WORK**
HOUR 13: _____ The hour is spent completing acid-base equations on paper. Students work individually on problems given in class. Problems and answers are provided in the lesson plans. **STUDENTS TURN IN WORK**	**HOUR 14:** _____ The copper-plating electrolysis experiment is explained. **HANDOUT PROVIDED** **RETURN STUDENT WORK**	**HOUR 15:** _____ The copper-plating electrolysis experiment is conducted. **NEED SPECIAL MATERIALS** **PREPARATION REQUIRED**
HOUR 16: _____ Review for a chemistry test.	**HOUR 17:** _____ Students take the "Chemistry Final Test." **TEST PROVIDED** **ANSWER SHEET PROVIDED**	**HOUR 18:** _____ Discussion of the test and of the chemistry unit in general. What did it provide, in addition to basic knowledge? **RETURN STUDENT WORK**

Lesson Plans and Notes

ATOMIC STRUCTURE (Hours 1, 2, 3, and 4)

HOUR 1: Introduce students to the study of chemistry: what it is and examples of how it is at work in our everyday lives. Students are asked if they can explain such things as the reaction between baking soda and vinegar or the fact that a small amount of salt "disappears" in a glass of water. A few simple demonstration experiments are helpful during this hour to whet students' appetites for chemistry. A general film about chemistry is also helpful.

HOUR 2: Give students the "Chemistry Lesson 1" handout that teaches about atomic structure. Spend the hour explaining the first three terms on the handout: "element," "atom," and "molecule."

Note:

- It is useful to have a variety of teaching aids, or props, to help students visualize atoms and molecules. A film or filmstrip can be shown, overheads might be used, and three-dimensional models of molecules are *very* effective. Showing samples of elements, compounds, and mixtures will help students see that atoms and molecules combine to form substances they recognize from the real world. For example, a piece of aluminum, a silver (or copper, or nickel) coin, a piece of gold jewelry, a diamond, a copper wire, and so forth, can illustrate what elements are. Substances such as salt, water, steel, brass, and plastic can illustrate compounds.

HOUR 3: Discuss the remaining terms from the handout on atomic structure: "proton," "electron," "neutron," "nucleus," "electron shell," and "octet." Show students how to draw atoms on paper, given the number of protons, neutrons, and electrons in each. Do examples on the board.

Note:

- The three-dimensional nature of atoms should not be emphasized in this course, although students should be told that the orbits of electrons are really more like spheres than circles. Tell your students that if they can understand what is happening in a single plane, they will have taken a giant step toward visualizing what really takes place in three dimensions.

HOUR 4: Students make concentric-circle representations of several atoms. Choose these from the periodic chart and give students this information for each element: number of protons, neutrons, and electrons. Come to class prepared with five or six problems, like the example below, for students to work on during this hour. Their drawings are handed in at the end of the hour.

Note:

• It is helpful to have compasses available during this hour, and future hours, when students make concentric-circle drawings.

Example: sulfur

    16 protons

    16 neutrons

    16 electrons

CONCENTRIC CIRCLE REPRESENTATION OF SULFUR

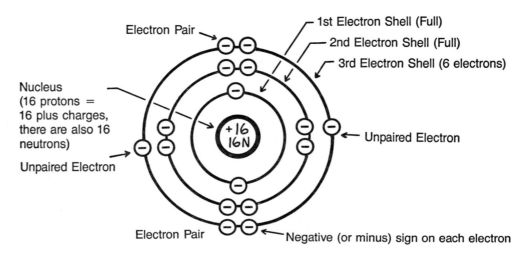

THE PERIODIC TABLE OF THE ELEMENTS (Hours 5, 6, 7, and 8)

HOUR 5: Return the drawings of atoms that the students did during Hour 4 and distribute the "Chemistry Lesson #2: The Periodic Table" handout. Explain the first two terms: "atomic number" and "atomic weight." Also distribute the "Periodic Table of the Elements" handout to students at the beginning of this hour; it shows the first 20 elements (and a few others that are interesting). Tell students to locate specific elements on their periodic tables and record how many protons, neutrons, and electrons an atom of each element will contain. Select these elements before class begins.

For the remainder of the hour students use their periodic table handout to search for elements, after being given only one piece of information about an atom. Once again, you are responsible for having these problems ready before the hour begins. For each problem, give the students only one clue:

1. How many protons an atom has, *or*

2. How many electrons an atom has, *or*

3. The total number of protons and neutrons an atom has

Note:

- Students should understand why there is a decimal place in the atomic weight of most elements. Explain that not all atoms of the same element weigh the same, because some have more or fewer neutrons than "normal." These abnormally heavy or light atoms are called isotopes and they make it necessary to give an *average* atomic weight. For the purposes of this course, *atomic weights should be rounded to the nearest whole number.* Also, explain the arrangement of elements on the periodic table by atomic number. From left to right, an atom of each succeeding element has one more proton and one more electron.

HOUR 6: Discuss density, using the "Chemistry Lesson 2" handout as a guide. Show the students how the density of elements can be obtained from the periodic table. The standard for measuring the density of solids and liquids is water, which has a density of 1. One cubic centimeter of water weighs one gram. Anything with a density greater than 1 will sink in water; a density less than 1 will float. Do a variety of problems, like the ones below, in class to help familiarize students with the concept of density. Define the units that are used to record density. Come to class prepared with at least ten problems, in addition to the examples provided. Collect student answers at the end of the hour.

Examples:

1. How much does 53 cc of boron weigh?
   (53 cc × 3.3 g/cc = 174.9 g of boron)
2. How much does 29 liters of fluorine weigh?
   (29 liters × 1.7 g/liter = 49.3 g of fluorine)
3. What volume of calcium weighs 131.2 g?
   (131.2 g ÷ 1.6 g/cc = 82 cc of calcium)
4. What volume of nitrogen weighs 128.7 g?
   (128.7 ÷ 1.3 g/liter = 99 liters of nitrogen)
5. What element has a density of 19.8 g/cc?
   (Plutonium)
6. Will lithium sink or float in water? How about boron?
   (Li floats; B sinks)

HOUR 7: Return the density problems students turned in last hour. On the board show students how to use information from the periodic table to draw representations of atoms of any element up to atomic number 20 (calcium). You may also want to discuss "groups" from the "Chemistry Lesson 2" handout. Have several sample drawings on the board when class begins; examples of such drawings are provided on the "Drawings of Atoms" handout which is given out this hour. Require students to come to the board, one at a time, to make drawings of their own, given a periodic table and the name of an element. Instructions to the student should be simple, such as: "John, come to the board and draw an atom of aluminum." The rest of the class is asked to draw these atoms on paper.

HOUR 8: Give students the names of several elements and instruct them to draw concentric-circle representations of these elements' atoms on paper at their desks. You may also ask the students to answer questions about the elements they have sketched.

Example: Magnesium

1. Draw an atom of magnesium.
2. How many neutrons does it have?
3. How many protons and electrons?
4. How much would 103 cc of Mg weigh?
5. Will magnesium float?

Be prepared with *at least* five such problems and have drawings made of each for reference. Discuss these problems at the end of the hour and require that student drawings be handed in.

HOUR 9: Return the students' drawings from Hour 8, and distribute the "Lesson 3: Ionic Bonding" and "Concentric-Circle Representation of the Water Molecule" handouts. Use this hour to explain the water molecule and its polarity. By the end of class students should understand that the water molecule is formed by *electron sharing,* and that this molecule has a partially negative side (the side opposite the two hydrogen atoms) and a partially positive side (the hydrogen atoms themselves). The electrons are *shared,* so the two hydrogen atoms and one oxygen atom form a molecule but do *not* become ions. The water molecule that is formed is called a *polar molecule,* because it has a positive *and* a negative charge. Its electrostatic strength (the attraction of its positive and negative side) is great enough to break the crystalline structure of many ionic compounds, which means they will *dissolve* in water.

HOUR 10: Give students the four ionization example handouts and show them how atoms of various elements gain or lose electrons to form ions. Spend the entire hour making drawings on the board to illustrate ionic reactions; first, illustrate how an atom with one outer-shell electron can lose its outer electron to an atom that has seven outer-shell electrons (Ionization Example 1: NaCl). Then show how an atom with two outer shell electrons can lose those electrons to an atom that has six outer electrons (Ionization Example 2: MgO); show how any two atoms, with one outer-shell electron each, can lose those electrons to an atom that has six outer electrons (Ionization Example 3: Li_2O); and how one atom that has two outer electrons can lose them to any two atoms that each have seven outer electrons (Ionization Example 4: $CaCl_2$).

Note:

- Each of the atom combinations described this hour completes an octet in one or more of the atoms and empties the outer shell of the other atom(s). This is what causes ions to be charged. An electron that leaves its orbit to fill another atom's octet takes its minus-one (-1) charge with it. When the electron is gone, the atom it came from is left with a plus-one $(+1)$ charge (one more proton than electrons). The same thing happens in the other atom, only in a negative sense: since it gains an electron it takes on a -1 charge. Electrostatic attractions between these "plus" and "minus" charges cause the ions to crystallize in their solid form; these crystals are called "salts" or salt crystals.

HOUR 11: Referring to their ionization examples handouts, students spend the hour making concentric-circle drawings of atoms and ions at their desks. Give them a list of ionic reactions to draw:

	Compound	*Ionic Reaction*
1.	NaCl (sodium chloride)	$Na + Cl \rightarrow NaCl \overset{H_2O}{\rightarrow} Na^{+1} + Cl^{-1}$
2.	LiFl (lithium fluoride)	$Li + Fl \rightarrow LiFl \overset{H_2O}{\rightarrow} Li^{+1} + Fl^{-1}$
3.	CaCl$_2$ (calcium chloride)	$Ca + 2Cl \rightarrow CaCl_2 \overset{H_2O}{\rightarrow} Ca^{+2} + 2Cl^{-1}$
4.	KFl (potassium fluoride)	$K + Fl \rightarrow KFl \overset{H_2O}{\rightarrow} K^{+1} + Fl^{-1}$
5.	AlCl$_3$ (aluminum chloride)	$Al + 3Cl \rightarrow AlCl_3 \overset{H_2O}{\rightarrow} Al^{+3} + 3Cl^{-1}$
6.	BeO (beryllium oxide)	$Be + O \rightarrow BeO \overset{H_2O}{\rightarrow} Be^{+2} + O^{-2}$

The assignment is to draw the three stages of each reaction:

1. Draw the atoms that make up the compound.
2. Draw the molecule and use dashed lines to mark where electron transfers will take place.
3. Draw the ions, showing all charges.

The answers are given below as "Lewis dot formulas," which show only the outer-shell electrons in each atom or ion. You should require students to do concentric-circle drawings, and collect their work at the end of the hour. An additional hour may be necessary to thoroughly explain ionization *and* ensure plenty of time for students to work on their drawings in class.

Lewis dot formula answers for this hour:

	ATOMS	$\rightarrow$	MOLECULES	$\overset{H_2O}{\rightarrow}$	IONS
1.	Na· + ·C̈l:	$\rightarrow$	Na :C̈l: $\overset{H_2O}{\rightarrow}$		Na^{+1} + :C̈l:$^{-1}$
2.	Li· + ·F̈l:	$\rightarrow$	Li :F̈l: $\overset{H_2O}{\rightarrow}$		Li^{+1} + :F̈l:$^{-1}$
3.	·Ca· + 2·C̈l:	$\rightarrow$	:C̈l: Ca :C̈l: $\overset{H_2O}{\rightarrow}$		Ca^{+2} + :C̈l:$^{-1}$ +:C̈l:$^{-1}$
4.	K· + ·F̈l:	$\rightarrow$	K:F̈l: $\overset{H_2O}{\rightarrow}$		K^{+1} + :F̈l:$^{-1}$
5.	·Al· + 3·C̈l:	$\rightarrow$	:C̈l: Al :C̈l: $\overset{H_2O}{\rightarrow}$:C̈l:		Al^{+3} + :C̈l:$^{-1}$+:C̈l:$^{-1}$ + :C̈l:$^{-1}$
6.	Be: + Ö:	$\rightarrow$	Be:Ö: $\overset{H_2O}{\rightarrow}$		Be^{+2} + :Ö:$^{-2}$

HOUR 12: Return the students' drawings from Hour 11 and distribute the "Acid-Base Reactions" handout. Teach students that, generally speaking, ionic substances that contain hydrogen ions (H^+) are *acids* (water is not acidic because it is not ionic), and ionic substances that contain hydroxide ions (OH^-) are *bases*. When an acid and a base are combined the result is an ionic solution. If the water was evaporated from this solution the result would be an ionic salt. Illustrate this on the board with concentric-circle drawings of hydrogen chloride + sodium hydroxide (HCl + NaOH). Students can also see this illustration on their "Acid-Base Reaction" handout. At the end of the hour show students how their knowledge of acids and bases allows them to write chemical equations. Use the following standard equation.

(*Acid + Base*) [H^+ ion] + [$-$ ion] + [$+$ ion] + [OH^- ion] $\rightarrow$ ionic salt + water

Example: (HCL + NaOH) [H^{+1}] + [Cl^{-1}] + [Na^{+1}] + [OH^{-1}] $\rightarrow$ NaCl + H_2O

Note:

• Explain that OH^- is an *ion* composed of two atoms that are sharing an electron. As a *unit,* the OH^- ion has one extra electron and therefore a charge of -1.

HOUR 13: Give students a list of acid-base equations to complete. You may also require a drawing or two. Students turn in their work at the end of the hour. Here is a sample list of equations for students to complete:

1. HCl + NaOH $\rightarrow$
2. HFl + KOH $\rightarrow$
3. HFl + LiOH $\rightarrow$
4. HCl + KOH $\rightarrow$
5. HCl + LiOH $\rightarrow$
6. HFl + NaOH $\rightarrow$
7. H_2S + 2NaOH $\rightarrow$
8. H_2S + 2KOH $\rightarrow$
9. H_2S + 2LiOH $\rightarrow$
10. 2HCl + $Be(OH)_2$ $\rightarrow$
11. 2HFl + $Mg(OH)_2$ $\rightarrow$
12. 2HCl + $Ca(OH)_2$ $\rightarrow$
13. 2HCl + $Mg(OH)_2$ $\rightarrow$
14. 2HFl + $Ca(OH)_2$ $\rightarrow$
15. 2HFl + $Be(OH)_2$ $\rightarrow$

An additional hour may be needed to demonstrate how to complete these equations; you may want to do them *all* in class or have students do them on the board or on drawing paper. Here are the answers to these equations:

1. HCl + NaOH $\rightarrow$ H^{+1} + Cl^{-1} + Na^{+1} + OH^{-1} $\rightarrow$ NaCl + H_2O
2. HFl + KOH $\rightarrow$ H^{+1} + Fl^{-1} + K^{+1} + OH^{-1} $\rightarrow$ KFl + H_2O
3. HFl + LiOH $\rightarrow$ H^{+1} + Fl^{-1} + Li^{+1} + OH^{-1} $\rightarrow$ LiFl + H_2O
4. HCl + KOH $\rightarrow$ H^{+1} + Cl^{-1} + K^{+1} + OH^{-1} $\rightarrow$ KCl + H_2O
5. HCl + LiOH $\rightarrow$ H^{+1} + Cl^{-1} + Li^{+1} + OH^{-1} $\rightarrow$ LiCl + H_2O

6. $HFl + NaOH \rightarrow H^{+1} + Fl^{-1} + Na^{+1} + OH^{-1} \rightarrow NaFl + H_2O$

7. $H_2S + 2NaOH \rightarrow 2H^{+1} + S^{-2} + 2Na^{+1} + 2OH^{-1} \rightarrow Na_2S + 2H_2O$

8. $H_2S + 2KOH \rightarrow 2H^{+1} + S^{-2} + 2K^{+1} + 2OH^{-1} \rightarrow K_2S + 2H_2O$

9. $H_2S + 2LiOH \rightarrow 2H^{+1} + S^{-2} + 2Li^{+1} + 2OH^{-1} \rightarrow Li_2S + 2H_2O$

10. $2HCl + Be(OH)_2 \rightarrow 2H^{+1} + 2Cl^{-1} + Be^{+2} + 2OH^{-1} \rightarrow BeCl_2 + 2H_2O$

11. $2HFl + Mg(OH)_2 \rightarrow 2H^{+1} + 2Fl^{-1} + Mg^{+2} + 2OH^{-1} \rightarrow MgFl_2 + 2H_2O$

12. $2HCl + Ca(OH)_2 \rightarrow 2H^{+1} + 2Cl^{-1} + Ca^{+2} + 2OH^{-1} \rightarrow CaCl_2 + 2H_2O$

13. $2HCl + Mg(OH)_2 \rightarrow 2H^{+1} + 2Cl^{-1} + Mg^{+2} + 2OH^{-1} \rightarrow MgCl_2 + 2H_2O$

14. $2HFl + Ca(OH)_2 \rightarrow 2H^{+1} + 2Fl^{-1} + Ca^{+2} + 2OH^{-1} \rightarrow CaFl_2 + 2H_2O$

15. $2HFl + Be(OH)_2 \rightarrow 2H^{+1} + 2Fl^{-1} + Be^{+2} + 2OH^{-1} \rightarrow BeFl_2 + 2H_2O$

HOUR 14: Return student work from Hour 13. Introduce the electrolysis experiment and explain the chemical reaction that takes place. Give students the "Chemistry: Electrolysis" handout which explains how a stainless steel cathode in a solution of copper sulfate is plated when current is passed into the cathode. Include these points in the discussion:

1. Electrons flow from the negative pole of the battery to the stainless steel cathode.

2. When put in water, copper sulfate ionizes, with copper becoming a positive ion.

3. The positive ions of copper are attracted to the extra electrons in the cathode (provided by the battery) and they attach themselves to it, thus plating it with copper.

4. The anode is made of copper. Every time a copper ion from the solution joins with two electrons (from the cathode) to become solid copper on the *cathode,* a copper atom on the anode gives up two electrons and goes into the solution as an ion. The electrons travel from the anode to the battery, and thus a flow of electrons is maintained between the negative and positive poles of the battery. The number of copper ions in the solution remains constant throughout the experiment. For each ion that attaches to the cathode, one releases from the anode.

HOUR 15: Conduct the electrolysis experiment in class and discuss what is observed. What happens if the wires are reversed on the battery? Is there a current flowing through the water?

Note:

- Most of the information and many of the concepts presented in this chemistry course are demonstrated in the electrolysis experiment. It is designed to be a demonstration *or* an individualized activity with students working in small groups. It can be used as an independent project, but only when adequate adult supervision is ensured. When conducting this experiment, these safety precautions must be followed:

a. Rubber gloves and safety goggles must be worn.

b. A dry cell battery provides sufficient power. If a power transformer is used, it must be 12 volts or less.

c. Only one student at a time may be in charge of an experiment; all others must keep well away from the table.

d. Chemicals and solutions (copper sulfate in this case) must never be put in the mouth or eyes. Hands must stay away from faces during the experiment. If copper sulfate is ingested, have the student dilute it by drinking water, and call poison control or a doctor for advice. If copper sulfate gets in the eyes, flush with warm water for at least 15 minutes and call a doctor.

e. All equipment must be cleaned before students remove their gloves or goggles.

f. After the experiment is completed students must thoroughly wash their hands and clean their rubber gloves.

HOUR 16: Review for a final chemistry test. Work problems in class, have students make drawings on the board, review definitions, and answer questions.

HOUR 17: Students take the "Chemistry Final Test."

HOUR 18: Return and discuss the test, especially problems that were commonly missed. Spend the rest of the hour discussing the value of a unit on chemistry. It obviously sets the stage for a high school course, but more important, it gives students *confidence* in their ability to understand a difficult subject. If time permits, discuss careers in chemistry as well as the importance of chemistry in basic training for specific career areas: medicine, engineering, geology, agriculture, physics, meteorology, and many others.

General Notes About the Chemistry Project:

• All of the basic information needed to teach this project is provided in "Chemistry Project: Concepts for the Teacher."

• This project was designed to teach young students about chemistry, a subject normally reserved for the last year or two of high school. There are many activities and "hands-on" projects that can be incorporated into the presentation of chemistry in addition to what is provided in this book. Some suggestions:

a. Large poster drawings of atoms, ions, and molecules.

b. Chalkboard competitions: speed and accuracy of drawings of molecules, atoms, ions, or reactions.

c. Molecular models using styrofoam balls connected by wooden dowel rods or soda straws.

d. Carefully planned and supervised chemistry experiments. Demonstrations are better than having students actually conduct experiments unless you have facilities to handle such activities. Safety precautions must be followed.

e. You can make a list of concepts or phenomena that were covered during the project and require students to produce presentations about them. For example: atomic weight, density, polar water molecule, how ions are formed (e.g., $Na + Cl \rightarrow Na^+ + Cl^-$), how acids react with bases, octet, and so forth. Each student would then plan a short lesson to explain his or her topic to the class: kids teaching kids.

- You may find that additional time is necessary to explain certain information to your students or to review previous lessons before going on to the next topic. Plan your schedule to allow a few extra hours if they are needed.

Name _____ Date _____

CHEMISTRY LESSON 1:
Atomic Structure

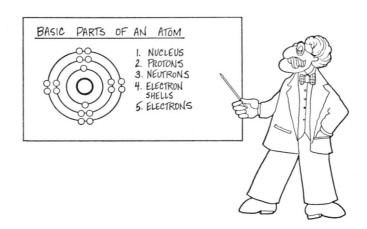

Students your age usually do not study chemistry. Some of the concepts of chemistry that you will learn in this course are not easy to understand, but if you pay attention and concentrate you can learn them as well as older kids do. Chemistry is a science that is learned in steps; you must understand one thing before you can understand the next. It is important that you take notes in class and study this handout carefully. If you do these two things you will do all right in this chemistry course. Below are a number of terms you will need to understand. They will be explained in class; you are *not* expected to read them by yourself and understand them perfectly. However, after they have been explained in class this handout should help you remember what they are and why they are important.

1. *Element:* Any substance that cannot be separated into simpler parts is called an element. Chocolate ice cream is not an element because it has a lot of different things in it. Water is made of oxygen and hydrogen, so water is not an element either. The *atoms* of an element must all be the same. Gold is an element because all gold atoms are the same. Other examples of elements are: iron, carbon, oxygen, sulfur, and hydrogen.

2. *Atom:* The smallest particle of an element is called an atom. An atom is *extremely* small, but it can still be called by the element's name. For example, if you have a billion atoms of gold, you have a small piece of gold; if you could cut off one atom from this you would *still* have gold. But if you could somehow cut this single gold atom in two you would no longer have gold, because one atom of gold is the smallest piece of gold that is possible. An atom is made up of protons, neutrons, and electrons. The protons and neutrons are found at the center of the atom, and this center is called the *nucleus.* The electrons orbit, or fly around, the nucleus. These orbits are called *electron shells.*

3. *Molecule:* When the atoms of more than one element combine they form a molecule. When two atoms of hydrogen combine with one atom of oxygen, a molecule of water is formed (H_2O). Every time you breathe you exhale millions of carbon dioxide molecules. A molecule of carbon dioxide is one atom of carbon combined with two atoms of oxygen (CO_2). Carbon dioxide and water are *not* elements because they each have more than one kind of atom.

4. *Proton:* Every atom is made up of three basic parts: protons, electrons, and neutrons. A proton is charged with one unit of positive electricity and is said to have a ($+1$) charge. Protons are found at the center of an atom, in the *nucleus.* A proton weighs almost 2,000 times more than an electron and about the same as a neutron.

5. *Electron:* The second basic part of every atom is the electron. An electron is charged with one unit of negative electricity, and is said to have a (-1) charge. Electrons orbit around the nucleus of an atom. They travel at close to the speed of light and have so little weight that they have no effect on the total weight of the atom.

6. *Neutron:* The third basic part of every atom is the neutron. A neutron has no electric charge and is said to have a (0) charge. Neutrons are found at the center of an atom, in the nucleus. Neutrons weigh the same as protons and their main importance is their contribution to the weight of atoms.

7. *Nucleus:* The center of an atom is called the nucleus. Protons and neutrons are packed into the nucleus and this is where an atom gets its *weight.* An atom is like a tiny solar system with the nucleus at the center and electrons in orbit around it.

8. *Electron shell:* Each electron has its own orbit around the nucleus, something like the planets around the sun. These orbits are all different distances from the center of the nucleus and each different orbit is called an electron shell. In the *first* electron shell there can be no more than two electrons. An atom that has more than *two* electrons will have more than one electron shell. In the *second* electron shell there can be no more than eight electrons. An atom that has more than *ten* electrons (two in the first shell and eight in the second) will have more than two electron shells. During this course you will learn how to draw atoms with their electron shells. *Be sure you take notes in class!*

9. *Octet:* Except for the first electron shell (which will take only two electrons), each electron shell around an atom would *like* to have eight electrons in it. This is called an *octet.* When the second shell has eight electrons it won't take any more, so additional electrons go to the third shell. When the third shell has eight electrons *it* won't take any more, so additional electrons go to the fourth shell. When an electron shell does not have eight electrons it "looks" for electrons in other atoms to help fill up its octet. This is how *molecules* and *ions* are formed. Once you understand what octets are, you can learn how atoms, molecules, and ions actually react with one another. Be sure to take notes as these concepts are explained in class.

CHEMISTRY LESSON 2:
The Periodic Table

THE PERIODIC TABLE TELLS THESE
THINGS ABOUT AN ELEMENT:

1. ATOMIC NUMBER
2. ATOMIC WEIGHT
3. DENSITY
4. WHICH "GROUP" IT
   BELONGS TO

1. *Atomic number:* An element's atomic number tells you how many protons one atom has in its nucleus *and* how many electrons it has in its electron shells. The number of protons and electrons in an atom is always the same, so if the atomic number of an element is eight you can immediately say that an atom of that element has eight protons and eight electrons.

2. *Atomic weight:* An element's atomic weight tells you the number of protons plus the number of neutrons an atom has in its nucleus. This is an average number because the number of neutrons can change slightly, but basically here is what you can tell from the atomic weight: if an element has an atomic weight of 31 you can say that the sum of protons and neutrons in one atom of that element equals 31. If you know that this element has an atomic number of 15 (15 protons) then you can tell how many neutrons there are by subtracting: $31 - 15 = 16$, or in other words, 15 protons + 16 neutrons = 31 (atomic weight). For this course, if the atomic weight is given with a decimal, round it to the nearest whole number before working with it.

3. *Density:* You can tell how "heavy" a substance is by knowing its density and its volume. You know that a handful of iron is heavier than a handful of feathers. This is because iron is *more dense* than feathers, or it has a greater *density*. In chemistry the density of solids and liquids is measured by how much one cubic centimeter of a substance weighs, in grams. One cubic centimeter of carbon weighs 2.3 grams, so the density of carbon is 2.3 *g/cc* (grams per cubic centimeter). If you have 10 cubic centimeters of carbon you can calculate its weight by multiplying 2.3 *g/cc* × (times) 10 *cc* = 23 g. Some elements are gases (oxygen, hydrogen, nitrogen, helium, neon and chlorine, for example). The density of a gas is measured by how much one *liter* of the gas weighs (at a specific temperature and pressure). One liter of oxygen weighs 1.4 grams, so the density of oxygen is 1.4 g/l (grams per liter). If you have 10 liters of oxygen you can calculate its weight by multiplying 1.4 g/l × 10 l = 14 g.

4. *Groups:* You can tell how many electrons an atom has in its *outer shell* by looking at the "group" its element is in. A "group" is a vertical column on the periodic table; these columns are labeled IA, IIA, IIIA, and so forth, up to VIIIA. Atoms in the IA group have *one* outer-shell electron. IIA atoms have *two* outer-shell electrons, IIIA atoms have *three* outer-shell electrons and so forth.

Name _____

Date _____

PERIODIC TABLE OF THE ELEMENTS
(Elements 1 Through 20)

IA	II A	III A	IV A	V A	VI A	VII A	VIII A
1 1.0 Hydrogen H (.1 g/l)							2 4.0 Helium He (.2 g/l)
3 6.9 Lithium Li (.5)	4 9.0 Beryllium Be (1.9)	5 10.8 Boron B (3.3)	6 12.0 Carbon C (2.3)	7 14.0 Nitrogen N (1.3 g/l)	8 16.0 Oxygen O (1.4 g/l)	9 19.0 Fluorine Fl (1.7 g/l)	10 20.2 Neon Ne (.9 g/l)
11 23.0 Sodium Na (1.0)	12 24.3 Magnesium Mg (1.7)	13 26.9 Aluminum Al (2.7)	14 28.1 Silicon Si (2.4)	15 31.0 Phosphorus P (1.8)	16 32.1 Sulfur S (2.1)	17 35.5 Chlorine Cl (3.2 g/l)	18 39.9 Argon Ar (1.8 g/l)
19 39.1 Potassium K (.9)	20 40.1 Calcium Ca (1.6)						

Some examples of heavy metals:

79 197.0 Gold Au (19.3)	82 207.2 Lead Pb (11.3)	94 244.0 Plutonium Pu (19.8)	47 107.9 Silver Ag (10.5)	92 238.0 Uranium U (19.7)

These five elements are not in order of increasing atomic number; they are arranged alphabetically and are included to show a few of the heavy metals.

Atomic Number → 6

Atomic Weight → 12.0

Element → Carbon

Symbol → C

Density → (2.3)

57

DRAWINGS OF ATOMS:
Concentric-Circle Representations

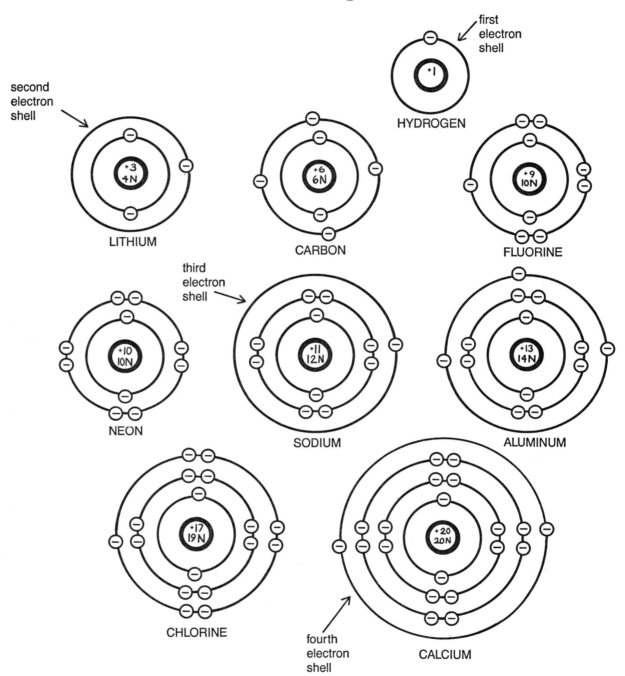

In each shell (except the first) electron pairs are formed after the shell contains four electrons. Carbon has four electrons in its outer shell, and therefore, *no* pairs. Neon has eight electrons in its outer shell; these eight electrons form *four* pairs.

CHEMISTRY LESSON 3:
Ionic Bonding

1. *Electron transfer:* Remember that an atom wants to have a full *octet* of electrons in its outer electron shell. Often atoms that need only one or two electrons to fill their octets (VIIA and VIA elements on the periodic table) will take electrons from other atoms to form "ions." Ions are atoms that have lost or gained one or more electrons (see the definition below). The process of giving up or gaining an electron is called an "electron transfer"; the electron "transfers," or goes, from one atom to another, forming a negative ion and a positive ion.

2. *Ion:* After an electron transfer occurs between two atoms, one atom is left with a positive (+) charge and the other atom gains a negative (−) charge. At least one electron (with a −1 charge) has moved from one atom to the other. An atom of sodium has one electron in its outer shell. An atom of chlorine has seven outer-shell electrons. When these two atoms come together, an electron transfer takes place: the sodium atom gives its single outer-shell electron to the chlorine. The resulting sodium ion has *eleven* protons and only *ten* electrons, so it has a plus one (+1) charge. The chlorine ion has *seventeen* protons and *eighteen* electrons, so it has a minus one (−1) charge. The sodium ion has *no* electrons in its outer shell. These ions of sodium and chlorine fit together in a neat pattern to form *salt crystals.* These crystals will dissolve in water. When this happens the ions become "free ions" in solution.

3. *Polar molecule:* Water can dissolve many ionic substances because it is a "polar molecule." This means that one side of the molecule has a positive, or plus, charge (+) and the other side has a negative, or minus, charge (−). The polar water molecule acts like a magnet and actually pulls certain ionic substances, like sodium chloride (table salt), apart. This is why table salt dissolves in water. After a substance dissolves in water, it is called a solution.

Concentric-Circle Representation
of the Water Molecule

The water molecule is not ionic. The electrons between the hydrogen atoms and the oxygen atom are *shared*. In ionic reactions the electrons are *transferred* from one atom to another, creating ions. The illustration below shows how two hydrogen atoms combine with one oxygen atom to form a polar water molecule. It is polar because a positive charge is created on one side of the molecule and a negative charge on the other. These charges allow water to dissolve ionic substances.

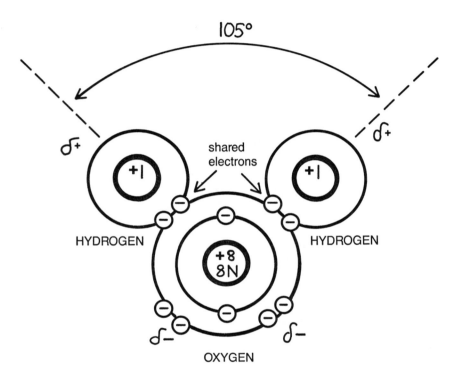

The symbol δ (delta) means "partial"; the water molecule has two extra electrons on one side and two extra protons on the other side. This gives the molecule two δ− charges on one side and two δ+ charges on the other side. These charges are what makes the water molecule *polar.*

Name _____ Date _____

IONIZATION EXAMPLE 1:
Sodium Chloride (NaCl)

THREE STAGES OF IONIZATION

1. Atoms Combine

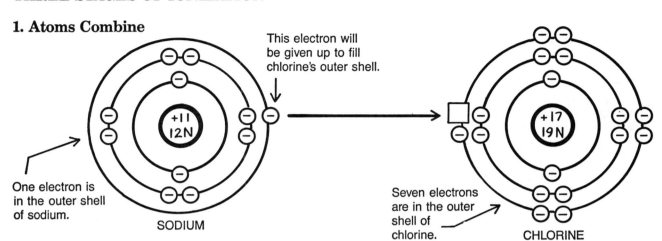

This electron will be given up to fill chlorine's outer shell.

One electron is in the outer shell of sodium.

SODIUM

Seven electrons are in the outer shell of chlorine.

CHLORINE

2. Electron Transfer

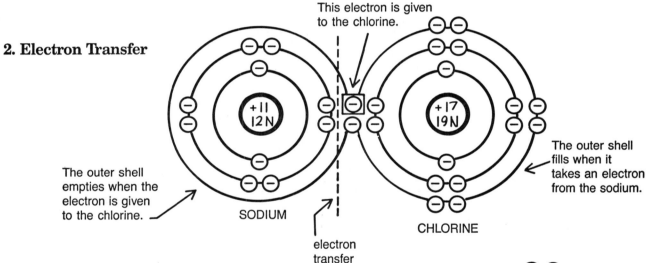

This electron is given to the chlorine.

The outer shell empties when the electron is given to the chlorine.

SODIUM

The outer shell fills when it takes an electron from the sodium.

CHLORINE

electron transfer

3. Ions Form

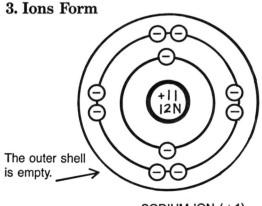

The outer shell is empty.

SODIUM ION (+1)

11 protons (+)
10 electrons (−)
+ 1 ionic charge

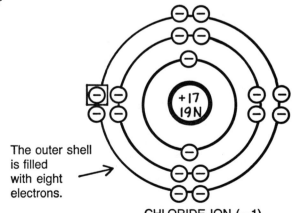

The outer shell is filled with eight electrons.

CHLORIDE ION (−1)

18 electrons (−)
17 protons (+)
−1 ionic charge

IONIZATION EXAMPLE 2:
Magnesium Oxide (MgO)

THREE STAGES OF IONIZATION

1. Atoms Combine

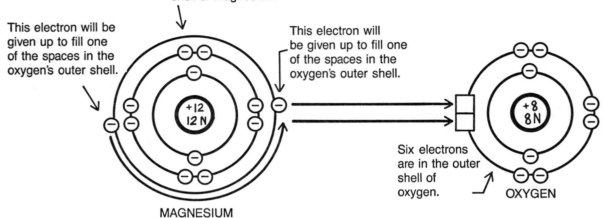

Two electrons are in the outer shell of magnesium.

This electron will be given up to fill one of the spaces in the oxygen's outer shell.

This electron will be given up to fill one of the spaces in the oxygen's outer shell.

Six electrons are in the outer shell of oxygen.

+12
12 N

+8
8 N

MAGNESIUM

OXYGEN

2. Electron Transfer

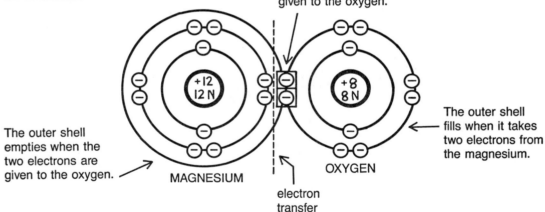

These two electrons are given to the oxygen.

The outer shell empties when the two electrons are given to the oxygen.

The outer shell fills when it takes two electrons from the magnesium.

+12
12 N

+8
8 N

MAGNESIUM

OXYGEN

electron transfer

3. Ions Form

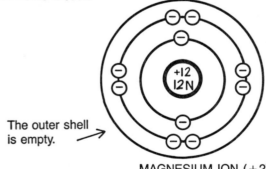

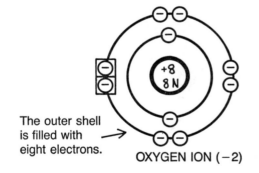

The outer shell is empty.

+12
12 N

MAGNESIUM ION (+2)

12 protons (+)
10 electrons (−)
+ 2 ionic charge

The outer shell is filled with eight electrons.

+8
8 N

OXYGEN ION (−2)

10 electrons (−)
8 protons (+)
− 2 ionic charge

IONIZATION EXAMPLE 3:
Lithium Oxide (Li₂O)

THREE STAGES OF IONIZATION

1. Atoms Combine

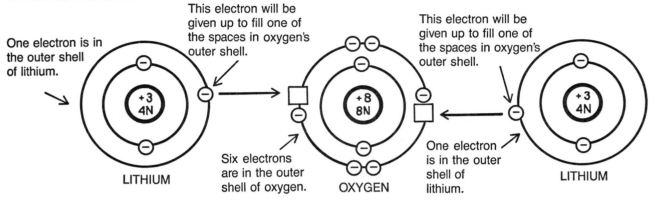

One electron is in the outer shell of lithium.

This electron will be given up to fill one of the spaces in oxygen's outer shell.

This electron will be given up to fill one of the spaces in oxygen's outer shell.

LITHIUM

Six electrons are in the outer shell of oxygen.

OXYGEN

One electron is in the outer shell of lithium.

LITHIUM

2. Electron Transfer

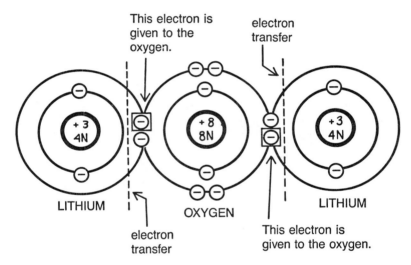

This electron is given to the oxygen.

electron transfer

LITHIUM

OXYGEN

LITHIUM

electron transfer

This electron is given to the oxygen.

3. Ions Form

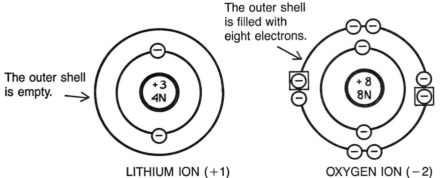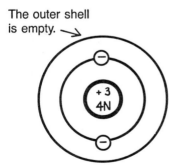

The outer shell is filled with eight electrons.

The outer shell is empty.

The outer shell is empty.

LITHIUM ION (+1)	OXYGEN ION (−2)	LITHIUM ION (+1)
3 protons (+)	10 electrons (−)	3 protons (+)
2 electrons (−)	8 protons (+)	2 electrons (−)
+1 ionic charge	− 2 ionic charge	+1 ionic charge

Name _____ Date _____

IONIZATION EXAMPLE 4:
Calcium Chloride (CaCl₂)

THREE STAGES OF IONIZATION

1. Atoms Combine

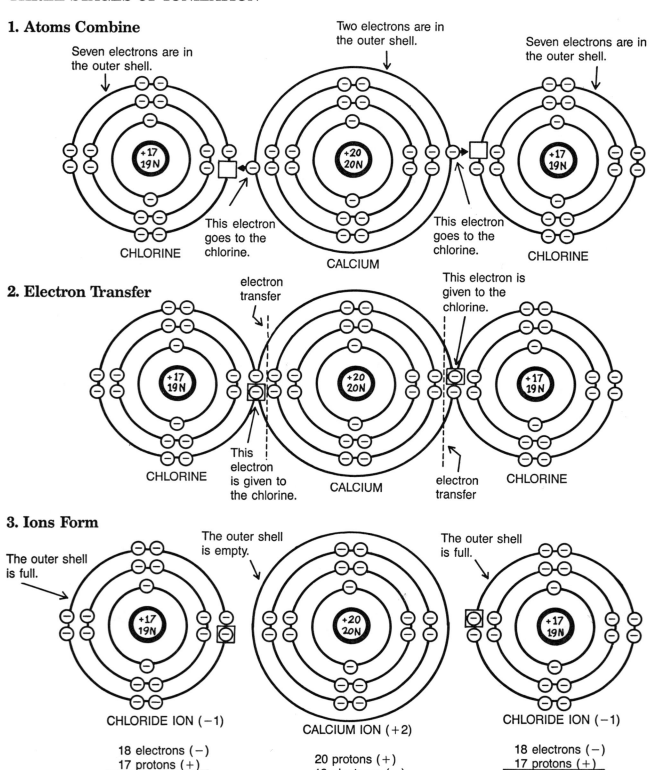

Seven electrons are in the outer shell.

Two electrons are in the outer shell.

Seven electrons are in the outer shell.

CHLORINE

This electron goes to the chlorine.

CALCIUM

This electron goes to the chlorine.

CHLORINE

2. Electron Transfer

electron transfer

This electron is given to the chlorine.

CHLORINE

This electron is given to the chlorine.

CALCIUM

electron transfer

CHLORINE

3. Ions Form

The outer shell is full.

The outer shell is empty.

The outer shell is full.

CHLORIDE ION (−1)

CALCIUM ION (+2)

CHLORIDE ION (−1)

| 18 electrons (−) |
| 17 protons (+) |
| − 1 ionic charge |

| 20 protons (+) |
| 18 electrons (−) |
| + 2 ionic charge |

| 18 electrons (−) |
| 17 protons (+) |
| − 1 ionic charge |

Name _____

Date _____

ACID-BASE REACTION:
Hydrogen Chloride + Sodium Hydroxide (HCl + NaOH)

THE THREE STAGES OF AN ACID-BASE REACTION

1. An acid (HCl) and a base (NaOH) are combined.

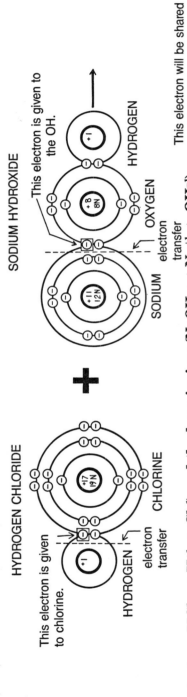

HYDROGEN CHLORIDE

This electron is given
to chlorine.

HYDROGEN

electron
transfer

CHLORINE

SODIUM HYDROXIDE

This electron is given to
the OH.

electron
transfer

SODIUM

OXYGEN

HYDROGEN

2. The acid ionizes (HCl→ + H⁺¹ + Cl⁻¹) and the base ionizes (NaOH→+ Na⁺¹ + OH⁻¹).

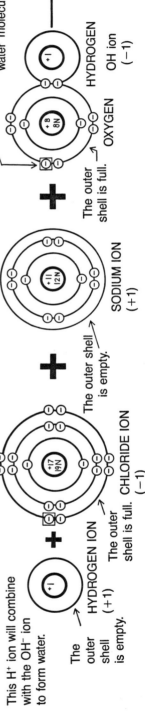

This H⁺ ion will combine
with the OH⁻ ion
to form water.

The
outer
shell
is empty.

HYDROGEN ION
(+1)

The outer
shell is full.

CHLORIDE ION
(−1)

The outer shell
is empty.

SODIUM ION
(+1)

This electron will be shared
with the H⁺ ion to form a
water molecule.

The outer
shell is full.

OXYGEN

HYDROGEN

OH ion
(−1)

3. The H⁺¹ ion from the acid combines with the OH⁻¹ from the base to form water. The other ions (Na⁺¹ and CL⁻¹) remain in solution.

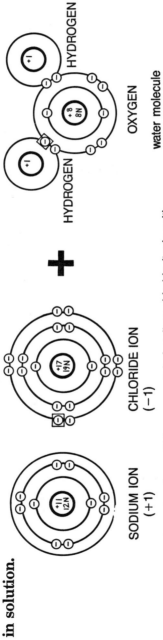

CHLORIDE ION
(−1)

SODIUM ION
(+1)

HYDROGEN

OXYGEN

HYDROGEN

water molecule

When not in solution, these two ions will form a crystal of sodium chloride (table salt).

INDIVIDUALIZED CLASSROOM EXPERIMENT:
Electrolysis

THE EXPERIMENT

This experiment will show how copper plating works. The procedure must be done with a teacher or other qualified adult supervising.

Materials Needed:

1. 1 liter of water
2. 2 tablespoons of copper sulfate
3. Mixing beaker (glass)
4. Stir rod
5. Dry cell battery
6. Copper anode
7. Stainless steel cathode
8. Copper wires to connect anode and cathode to the battery
9. Rubber gloves
10. Goggles
11. Paper towels and sponges
12. Safe place to conduct the experiment
13. Sink (for cleaning up)
14. Electrical outlet (if a transformer is used instead of a battery)

After the materials are collected, follow this procedure:

1. Make a 1-liter solution of water and copper sulfate. Use about 2 tablespoons of copper sulfate per liter of water.
2. Attach the copper anode to the positive pole of a dry cell battery.
3. Attach the stainless steel cathode to the negative pole of the dry cell.
4. The cathode will begin to get dark as copper ions come out of solution and attach themselves to the stainless steel. This is called "copper plating."
5. The anode will begin to deteriorate as copper atoms give up their electrons to the positive pole of the battery and go into solution as ions.

Safety Precautions:

1. You must have an adult supervisor with you when you conduct this experiment.
2. Wear rubber gloves and goggles.
3. Do not put your hands to your face. If you must scratch your nose, ask the supervisor to help remove one of your rubber gloves.
4. Great care must be taken when working with chemicals. Do not splash or spill anything. If something is spilled accidentally, wipe it up immediately with wet paper towels.
5. Since this experiment involves working with chemicals and electricity it is crucial that it be carefully conducted. Only one student may be in charge of an experiment at a time. All other students must stand back and observe quietly.
6. Clean all equipment before removing your gloves and goggles.
7. Thoroughly clean your rubber gloves and then wash your hands.

Optional Assignment:

1. Set up the electrolysis experiment as a demonstration.
2. Using posters and diagrams, explain:
   a) This chemical equation:

   $$CuSO_4 + H_2O \rightarrow Cu^{++} + SO_4^{--} + H_2O$$

   b) What a copper atom and a copper ion look like.
   c) Why the electrolysis experiment works.

ELECTROLYSIS SETUP

Electricity can be used to copper plate certain metal objects, as is shown in the illustration here. Study this page and try to understand what is happening. Be *sure* to ask questions about anything you are unsure of.

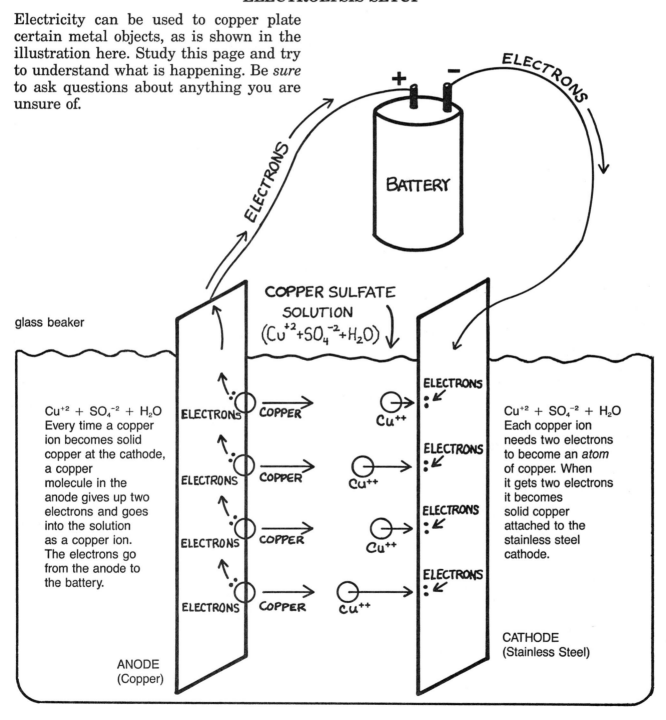

glass beaker

COPPER SULFATE SOLUTION $(Cu^{+2} + SO_4^{-2} + H_2O)$

$Cu^{+2} + SO_4^{-2} + H_2O$
Every time a copper ion becomes solid copper at the cathode, a copper molecule in the anode gives up two electrons and goes into the solution as a copper ion. The electrons go from the anode to the battery.

$Cu^{+2} + SO_4^{-2} + H_2O$
Each copper ion needs two electrons to become an *atom* of copper. When it gets two electrons it becomes solid copper attached to the stainless steel cathode.

ANODE
(Copper)

CATHODE
(Stainless Steel)

BATTERY

Name _____ Date _____

CHEMISTRY FINAL TEST

DIRECTIONS: You may use your class notes and the handouts that were used during this course to take this test. Be neat, label all of your drawings, and think each question through carefully. Check your answers before handing in the test. Do all of the drawings on separate paper and be sure your name is on everything.

1. How many protons are there in an atom of sulfur? _____

2. How many electrons are there in an atom of aluminum? _____

3. How many neutrons are there in an atom of beryllium? _____

4. How many electrons are in the *outer shell* of an atom of fluorine? _____

5. How much does 28 cubic centimeters of gold weigh? _____

6. How many neutrons are there in an atom of magnesium? _____

7. How many electrons are there in an atom of potassium? _____

8. How many protons are there in an atom of silver? _____

9. How many electrons are in the *outer shell* of an atom of sodium? _____

10. How much does 96 cubic centimeters of phosphorus weigh? _____

11. How many hydrogen atoms will attach themselves to an atom of carbon? _____

12. Draw an atom of silicon.

13. Draw an atom of beryllium.

14. Draw the electron transfer (transition stage) of $CaCl_2$ in this equation:
    $Ca + 2 Cl \rightarrow Ca^{+2} + 2 Cl^{-1}$.

15. Draw sodium fluoride (NaFl) in its three stages as it goes from atoms to ions:
    $Na + Fl \rightarrow Na^{+1} + Fl^{-1}$.

    a) Draw the separate atoms.
    b) Draw the transition stage (show electron transfer).
    c) Draw the ions (show all charges).

16. Complete the following acid-base reactions.

    a) $HCl + NaOH \rightarrow$
    b) $HFl + LiOH \rightarrow$
    c) $H_2S + 2NaOH \rightarrow$
    d) $2HCl + Mg(OH)_2 \rightarrow$
    e) $2HFl + Ca(OH)_2 \rightarrow$

Extra Credit

1. Finish this equation: $3HFl + Al(OH)_3 \rightarrow$
2. You will get two substances on the right side of the final equation. One of them is water and the other is a salt.

    a) Draw the atoms that make up the salt.
    b) Draw the atoms in their transition stage (when electrons are being transferred).
    c) Draw the ions, showing all charges.

ANSWERS TO CHEMISTRY FINAL TEST

1. sixteen
2. thirteen
3. five
4. seven
5. 540.4 g
6. twelve
7. nineteen
8. forty-seven
9. one
10. 172.8g
11. four
12. silicon:

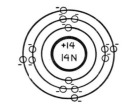

13. beryllium:

14. $CaCl_2$:

electron transfer

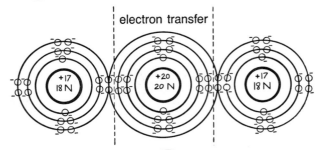

15. a) Na and Fl atoms:

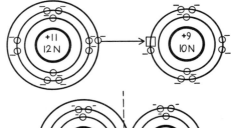

b) NaFl electron transfer:

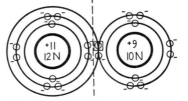

c) $Na^{+1} + Fl^{-1}$ ions:

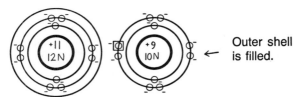

No electrons are in the outer shell. →

Outer shell is filled.

16. a) $HCl + NaOH \rightarrow H^{+1} + Cl^{-1} + Na^{+1} + OH^{-1} \rightarrow NaCl + H_2O$

   b) $HFl + LiOH \rightarrow H^{+1} + Fl^{-1} + Li^{+1} + OH^{-1} \rightarrow LiFl + H_2O$

   c) $H_2S + 2NaOH \rightarrow 2H^{+1} + S^{-2} + 2Na^{+1} + 2OH^{-1} \rightarrow Na_2S + 2H_2O$

   d) $2HCl + Mg(OH)_2 \rightarrow 2H^{+1} + 2Cl^{-1} + Mg^{+2} + 2OH^{-1} \rightarrow MgCl_2 + 2H_2O$

   e) $2HFl + Ca(OH)_2 \rightarrow 2H^{+1} + 2Fl^{-1} + Ca^{+2} + 2OH^{-1} \rightarrow CaFl_2 + 2H_2O$

Extra Credit:

1. $3\,HFl + Al(OH)_3 \rightarrow 3H^{+1} + 3Fl^{-1} + Al^{+3} + 3OH^{-1} \rightarrow AlFl_3 + 3H_2O$

2a.

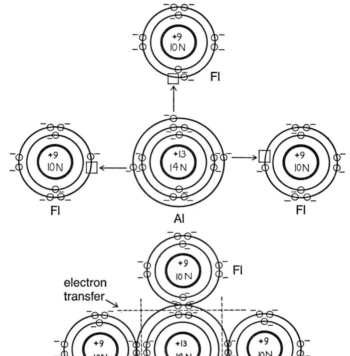

2b.

electron transfer

electron transfer   electron transfer

2c.

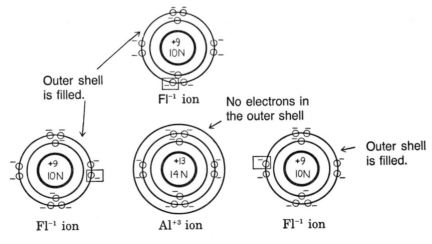

Outer shell is filled.

Fl^{-1} ion

No electrons in the outer shell

Outer shell is filled.

Fl^{-1} ion Al^{+3} ion Fl^{-1} ion

ANIMAL TAXONOMY

Teacher Preview

General Explanation:
This project provides a solid, basic outline of information about the five classes of vertebrate animals. It lays the groundwork for extended animal study by giving students necessary background knowledge.

Length of Project: 9 hours

Level of Independence: Basic

Goals:

1. To prepare students to work on future animal-related projects.
2. To introduce students to a basic science discipline (zoology).
3. To emphasize notetaking, listening, concentration, and other study skills.
4. To teach students basic classification information about vertebrate animals.

During This Project Students Will:

1. Be introduced to a system of classifying vertebrate animals.
2. Study animal adaptations: what they are and how they are used.
3. Be introduced to a body of information about each of the five classes of vertebrate animals: fish, amphibians, reptiles, birds, and mammals.
4. Analyze which class almost any vertebrate animal belongs in, given adequate information about its breathing system, heating system, method of reproduction, and body covering.
5. Take notes in class.

Skills:

Collecting data	Organizing
Listening	Outlining
Observing	Concentration
Summarizing	Writing
Handwriting	Spelling
Neatness	

Handouts Provided:

- "Animal Terms and Definitions"
- "Animal Taxonomy Final Test" (answers provided)

PROJECT CALENDAR:

HOUR 1: _____

Introduction to the study of animals and discussion of the term "classification."

HOUR 2: _____

Discussion of the term "adaptation" as it relates to the animal world. Adaptations are described for several animals. Students are given the "Animal Terms and Definitions" handout.

PREPARATION REQUIRED
HANDOUT PROVIDED

HOUR 3: _____

Introduction to the five classes of vertebrate animals. A method of identifying each class is described.

HOUR 4: _____

Lecture titled "Taxonomy of Vertebrate Animals" is begun: FISH and AMPHIBIANS.

PREPARATION REQUIRED

HOUR 5: _____

Lecture: REPTILES

PREPARATION REQUIRED

HOUR 6: _____

Lecture: BIRDS

PREPARATION REQUIRED

HOUR 7: _____

Lecture: MAMMALS

PREPARATION REQUIRED

HOUR 8: _____

Review for final test.

HOUR 9: _____

Final test.

Lesson Plans and Notes

HOUR 1: Introduce students to the study of animals. Some useful ways to do this would be: a film, filmstrip, trip to the zoo or nature center, classroom presentation by a naturalist, zoologist, zoo volunteer, or outdoor educator, or any other method of directing students' attention to animals. Students discuss why animals are important to the world and how people go about studying them. Discuss the need to *identify* animals before they can be studied. This leads to the word "classification." Put the following outline for animal classification, arranged from most general to most specific, on the board for students to copy:

I. Kingdom: Animal

  A. Phylum: Chordates (animals with a central nervous system)

    1. Subphylum: Vertebrates (animals with a backbone)
      a) Class: fish, amphibian, reptile, bird, or mammal

        (1) Order: There are several orders for each class

          (a) Family: There are several families for each order

            i) Genus: There are several genera for each family

            ii) Species: There are several species for each genus

The African lion, for example, is outlined like this:

AFRICAN LION

I. Kingdom: Animal

  A. Phylum: Chordata

    1. Subphylum: Vertebrata
      a) Class: Mammalia

        (1) Order: Carnivora

          (a) Family: Felidae

            i) Genus: Felis

            ii) Species: Leo

Notes:

- Be sure students understand that they will not be required to learn Latin for this course. The outline shows how classification is done, but this study of animal classification covers kingdom through order only. Family, genus, and species are rarely talked about, and only a relatively few orders are examined. Primary emphasis is placed on the five classes of vertebrate animals.

- The definition for "chordates" is more complicated than just "an animal with a central nervous system." Technically it should be an animal which, at some point in its development, has each of these three characteristics:

a. A notochord,

b. A dorsally located central nervous system and

c. Gill clefts.

If you wish to teach it this way, insert it into the outline. There are animals that have central nervous systems but that are not chordates.

- If the introduction to animals requires an entire hour, an hour should be added for outlining the system of animal classification that is provided.

HOUR 2: Give students the "Animal Terms and Definitions" handout for reference throughout the project. Spend this hour discussing and illustrating animal adaptations. Show students a picture of an animal, and ask them to identify the adaptations that animal has to help it survive in its natural environment. Ask questions about the animal to help students think of things, for example: "What does the rattlesnake have that helps it eat, move, hide, defend itself, and survive under different conditions?"

Here is a typical list of animals to study for adaptations:

1. Giraffe (mammal)
2. Skunk (mammal)
3. Boa constrictor (reptile)
4. Elephant (mammal)
5. Owl (bird)
6. Chameleon (reptile)
7. Eagle (bird)
8. Frog (amphibian)
9. Turtle (reptile)
10. Tiger (mammal)
11. Rattlesnake (reptile)
12. Shark (fish)
13. Hummingbird (bird)
14. Piranha (fish)

Note:

- It is important for the rest of the course that students understand adaptations. Spend another hour discussing and illustrating them if necessary. Be sure you are well prepared with pictures and notes before beginning this hour.

HOUR 3: Students are introduced to the five classes of vertebrate animals. Ask students to identify common animals that fit within each of the five classes. These classes are presented from the simplest to the most complex:

1. Fish
2. Amphibians
3. Reptiles
4. Birds
5. Mammals

The animals in these classes are progressively

1. less dependent upon water
2. more complicated in their central nervous systems
3. more able to regulate their body temperature (through mobility as well as physical structure)

There are four major characteristics that are used to identify what class an animal comes from. These are

1. Body covering
2. Respiration system
3. Body heating system
4. Reproduction system

Students are introduced to the terms "sub" and "super" as they are used in classifying animals (see the handout "Animal Terms and Definitions"). There are subphylums and superclasses and subclasses. This can be confusing and students should be aware that such terms are used because the animal world does not fit perfectly into any pattern, and methods of classifying animals have to be flexible enough to accommodate eccentricities of nature and deviations from the norm.

HOUR 4: Begin an outline of the five classes of animals this hour and continue it for approximately four hours of instruction. Prepare for Hours 4-7 by thoroughly understanding the material in the outline. It is also very helpful to have visual aids available. Students can be required to take notes.

Taxonomy of Vertebrate Animals

I. FISH

  A. Four major characteristics:

    1. Body covering—moist, slimy scales (with a few exceptions, like the bullhead, which have no scales)
    2. Heating system—ectothermic (cold-blooded)
    3. Respiration system—breathe with gills only
    4. Reproduction system—usually eggs laid in water (exceptions: guppies and a few others give live birth)

  B. There are two "superclasses" of fish:

    1. Fish with jaws
    2. Fish without jaws

  This course is concerned with only the superclass: *fish with jaws.*

  C. This superclass is further divided into two classes of fish with jaws, and each of these classes is divided into two orders:

    1. Cartilaginous fishes (class)

      a) sharks (order)
      b) rays (order)

    2. Bony fishes (class)

      a) higher bony fishes (order)
      b) lung fish (order)

  D. Other facts about fish

    1. There are 20,000 species of fish. These 20,000 species of fish represent approximately 40 percent of all living vertebrate species in the world.

2. Fish have a primitive two-chambered heart.
3. Fish have no external ears, but do have internal ears that help with balance in the water.
4. Most fish can smell with their nostrils (nares).
5. Fish have "swim bladders" that help hold them at different levels in the water.
6. Fish have "lateral lines" on their sides to help detect vibrations and changes in pressure.

II. AMPHIBIANS

A. Four major characteristics:
1. Body covering: moist smooth skin
2. Heating system: ectothermic
3. Respiration system: breathe with gills *and* lungs at different stages in life
4. Reproduction system: jelly-like eggs, usually laid in water

B. There are three orders of amphibians:
1. Salamanders and newts
2. Frogs and toads
3. Caecilians (earthworm shaped)

C. Other facts about amphibians:
1. Some respiration occurs through the skin during the transition from gill-breather to lung-breather.
2. Eggs must be laid where it is moist; the eggs have no shell to protect from dehydration.

HOUR 5: Continue lecture on five animal classes.

III. REPTILES

A. Four major characteristics
1. Body covering: dry scales
2. Heating system: ectothermic
3. Respiration system: breathe with lungs only
4. Reproduction system: leathery-shelled eggs laid on land

B. There are four orders of reptiles:
1. Snakes and lizards (squamata)
2. Turtles (testudines)
3. Crocodilians: crocodiles, alligators, caimens, gavials (loricata)
4. Tuatara

C. Other facts about reptiles:
1. Snakes *do* have backbones.
2. Snakes "feel" and "taste" the air by flicking their tongues in and out.
3. Snakes have no eyelids.
4. Reptiles may be aquatic, terrestrial, or arboreal.

5. The top shell of a turtle is the "carapace." The bottom shell is the "plastron."
6. There are lizards that have no legs.
7. There are two physiological differences between snakes and lizards:

   a) Lizards have eyelids and snakes do not.

   b) Lizards have ears and snakes do not.

HOUR 6: Continue lecture on five animal classes.

IV. BIRDS

A. Four major characteristics:
   1. Body covering: feathers
   2. Heating system: endothermic (warm-blooded)
   3. Respiration system: breathes with lungs only
   4. Reproduction system: brittle-shelled eggs

B. There are 27 orders of birds.

C. There are 155 living families of birds.

D. There are over 8,600 living species of birds.

E. Most birds can be placed in one of seven "lifestyle" categories:
   1. Waterfowl: ducks, geese, swans, grebes, loons, coots, penguins, auks
   2. Climbers: macaws, parrots, parakeets, lorikeets, woodpeckers, nuthatches
   3. Perchers: chickens, turkeys, bluejays, pigeons, sparrows, robins, cardinals
   4. Raptors: eagles, hawks, owls, falcons
   5. Scavengers: vultures, crows, cara caras
   6. Waders: flamingos, cranes, herons, egrets, ibises
   7. Grazers: ostriches, emus, rheas, kiwis

F. Other facts about birds:
   1. Birds have three kinds of feathers: contour, flight, and down.
   2. Birds molt.
   3. Most birds have a preen gland.
   4. Some birds migrate.
   5. Birds have a four-chambered heart.
   6. Most birds have very good binocular and peripheral vision.
   7. Birds must be light to fly:

      a) No teeth (food is ground in the gizzard and crop)
      b) Feathers instead of fur
      c) Hollow bones
      d) Very small stomach

   8. Six examples of beak and feeding adaptations:

      a) Nectar syringe (hummingbird)
      b) Soup strainer (flamingo)

c) Seed cracker (cardinal)
d) Fruit/nut cracker (parrot)
e) All-purpose (crow)
f) Knife/hook (hawk)

9. Six examples of foot adaptations:

a) Perching (robin)
b) Walking (ostrich)
c) Climbing (woodpecker)
d) Wading (crane)
e) Swimming (duck)
f) Tearing (eagle)

HOUR 7: Continue lecture on five animal classes.

V. MAMMALS

A. Four major characteristics:

1. Body covering: hair or fur at some time in life
2. Heating system: endothermic (warm-blooded)
3. Respiration system: breathe with lungs only
4. Reproduction: live birth (with two exceptions: the two monotremes)

B. There are three subclasses of mammals:

1. Monotremes (platypus and spiny anteater)
2. Marsupials (koala, kangaroo, opossum)
3. Placentals (almost all mammals)

a) There are seventeen orders of placental mammals. Six examples are

(1) Rodentia—mice, squirrels, rats, beavers, gophers
(2) Artiodactyla (hoofed animals with an even number of toes)—pigs, goats, deer, cattle, antelope, hippos
(3) Perissodactyla (hoofed animals with an odd number of toes)—horses, zebras, rhinos
(4) Cetacea—whales, porpoises, narwhals
(5) Carnivora—cats, dogs, weasels, bears
(6) Primates—tree shrews, monkeys, apes, humans

b) Placental mammals live almost everywhere on the earth.
c) Placentals are the most developed of the three subclasses of mammals.

C. Other facts about mammals. Mammals have

1. Milk glands in females (mammary glands)
2. Well-developed brain and nervous system, particularly the cerebral hemisphere (area for memory and reason)
3. Teeth socketed in jaw bones
4. Well-developed face muscles
5. External ear openings

6. Larynx (vocal cords)

7. Fleshy lips

D. Two major groupings of primates:

1. Prosimians

  a) Tree shrews (most primitive)

  b) Lemurs, tarsiers, aye ayes

2. Higher primates

  a) Monkeys

   (1) Old world (Africa and Asia) and new world (North and South America)

   (2) More developed than prosimians

  b) Anthropoid apes

   (1) Gorilla (mountain and lowland), chimpanzee, orangutan, gibbon

   (2) The main differences between monkeys and apes:

    (a) Apes have more developed brains

    (b) Monkeys cannot flex their shoulders or put both arms about their head

  c) Human beings

   (1) Most developed creature on earth

HOUR 8: Review session for the final test.

HOUR 9: Final test.

General Notes About This Project:

- The final test is fairly difficult. To make it easier, you may want to give students a sheet containing all of the answers listed in the wrong order, thus making it into a gigantic matching test. Your judgment will be necessary to ensure a test that is challenging but not overwhelming. You may also want to allow students to use their notes during the test: this is a good way to "test" the quality of their note taking.

- In its most basic form, this project can teach outlining and provide background information for your students to use in future projects. You need not understand all of the terms or be familiar with all of the animals mentioned in the outline to make the material useful. Similarly, students need not learn all of the facts in the outline to benefit from the note taking and outlining skills that are emphasized.

- There are several animal study projects which follow "Animal Taxonomy" in this book. The nine or so hours that are devoted to giving students animal classification background should be viewed as an investment in future projects.

- If you are going to teach outlining and note taking with this project, the material should be presented, in outline form, from the front of the room. A basic outline can be written on the board and the rest of the material presented orally. This method of giving information to students is slower than giving them handouts, but it places emphasis on two very important academic skills: taking notes in class and outlining information.

• This project can be enhanced by planning ahead and ordering a variety of films, film strips, magazines, posters, and books that complement the outlined information. The use of these materials may require additional hours of instruction time. Specimens, models, displays, charts, posters, diagrams, slides, and other visual materials may be available from

1. High school science departments
2. Local college biology departments
3. Zoos
4. Nature centers
5. Museums
6. Nature societies

• If possible, plan to visit a zoo or nature center as a culminating activity to this project. Have students observe animals for adaptations and classification characteristics.

Name _____ Date _____

ANIMAL TERMS AND DEFINITIONS

Adaptation: anything an animal has or does that helps it survive in its natural surroundings.

Aquatic: living in water or found naturally in water.

Arboreal: living in trees or found naturally in trees.

Binocular vision: animals with binocular vision have eyes in the front of the head that give depth perception.

Carnivore: an animal that eats meat.

Cartilaginous: made of cartilage instead of bone; one of the two superclasses of fish.

Characteristics: physical traits that can be used to identify an animal: how it looks and acts.

Class: one of the five major divisions of vertebrate animals: fish, amphibians, reptiles, birds, and mammals.

Dehydrate: to lose water, evaporate, or dry up.

Diurnal: active during the day.

Ectothermic: cold-blooded; not able to internally control or regulate body temperature.

Endothermic: warm-blooded; able to control or regulate body temperature internally (by generating body heat).

Food web: the relationship, in terms of diet, amongst the living things in a given area; what each animal and plant eats. A complete food web shows every source of food for every animal in the area.

Herbivore: an animal that eats plants.

Larynx: vocal cord found in all mammals.

Lateral line: organ on fish that helps detect pressure and vibrations.

Mammary gland: milk gland found on all female mammals.

Migrate: the seasonal movement of animals to different territories. This is done primarily to find food or for mating purposes.

Molt: birds periodically lose their worn-out feathers to make room for new replacement feathers.

Nares: the nostrils of a fish; fish can smell with their nares.

Natural habitat: the type of territory, or surroundings that a particular animal lives in in the wild if there are no outside influences.

Nocturnal: active at night.

Omnivore: an animal that eats plants and meat.

Peripheral vision: animals with peripheral vision have eyes that are able to see to both sides, without turning their heads.

Placental mammal: an animal from the largest and most developed subclass of mammals: a fetus in a placental mammal is carried in the uterus until it is fully developed. The umbilical cord is attached to the *placenta,* which feeds the fetus until birth. Placental mammals give live birth.

Population: the number of animals living within a certain area at any given time (such as the wolf population of Alaska in 1985).

Preen gland: gland in many birds that supplies oil for waterproofing and insulating feathers.

Primate: the most-developed order of mammals; it includes monkeys, apes, and human beings.

Reproduction: system by which animals have their young (eggs in water, eggs on land, live birth).

Respiration: system used by animals to breathe (gills, lungs, or both).

Subclass: a term indicating that there is a major division just below a class and above an order.

Superclass: a term indicating that there is a major division just above a class and below a subphylum.

Swim bladder: organ in fish that helps them hold steady at different levels and keeps them from sinking.

Terrestrial: living on land or found naturally on land.

Territory: the area of land an animal or group of animals claims as its own and is willing to defend.

Vertebrate: an animal with a backbone (or spinal column).

Name _____ Date _____

ANIMAL TAXONOMY FINAL TEST

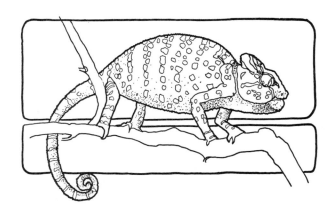

1. Number the classification terms below from 1 to 8; 1 is the biggest group of animals possible and 8 is the smallest group possible when classifying animals.

_____ order   _____ species   _____ kingdom   _____ class

_____ subphylum   _____ genus   _____ family   _____ phylum

2. What are the four major characteristics of fish?

a) _____

b) _____

c) _____

d) _____

3. There are two "superclasses" of fish. Which one are we studying?

_____

4. This "superclass" of fish has two classes, and each of these classes has two orders. Below are six terms for these classes and orders. Write them correctly into the outline.

rays, bony fishes, lungfish, cartilaginous fishes, higher bony fishes, sharks

a) Class: _____

   (1) Order: _____

   (2) Order: _____

b) Class: _____

   (1) Order: _____

   (2) Order: _____

ANIMAL TAXONOMY FINAL TEST (continued)

5. What kind of heart does a fish have?

_____

6. What are "nares"?

_____

7. What do fish have that helps hold them at different levels in the water?_____

8. What do fish have on their sides to help them detect vibrations and changes in

water pressure?_____

9. What are the four major characteristics of amphibians?

a) _____

b) _____

c) _____

d) _____

10. Name two of the three orders of amphibians.

a) _____

b) _____

11. When does an amphibian "breathe" through its skin?

_____

12. Why must an amphibian's eggs be laid where it is moist?

_____

13. What are the four major characteristics of reptiles?

a) _____

b) _____

c) _____

d) _____

14. Name three of the four orders of reptiles.

a) _____

b) _____

c) _____

Name _____ Date _____

ANIMAL TAXONOMY FINAL TEST (continued)

15. Do snakes have backbones?_____

16. Why do snakes flick their tongues out?

_____

17. Do snakes have eyelids?_____

18. What are the top and bottom of a turtle's shell called?

   a) Top: _____

   b) Bottom: _____

19. What are the two physical differences between lizards and snakes?

   a) _____

   b) _____

20. What are the four major characteristics of birds?

   a) _____

   b) _____

   c) _____

   d) _____

21. How many orders of birds are there? (circle the correct answer)

   6   15   27   68

22. Write two examples for each category of bird lifestyle listed below. The first one is done for you.

	A	B
a) Waterfowl	ducks	geese
b) Climbers	_____	_____
c) Perchers	_____	_____
d) Raptors	_____	_____
e) Scavengers	_____	_____
f) Waders	_____	_____
g) Grazers	_____	_____

Name _____ Date _____

ANIMAL TAXONOMY FINAL TEST (continued)

23. What three kinds of feathers do birds have?

a) _____

b) _____

c) _____

24. Birds_____(lose their feathers) once or twice a year.

25. What is a preen gland used for? _____

_____

26. What kind of heart does a bird have? _____

27. Name three adaptations a bird has that helps it be light enough to fly.

a) _____

b) _____

c) _____

28. Write one example for each beak and feeding adaptation listed below. The first one is done for you.

a) Seed cracker: _____cardinal_____

b) Nectar syringe: _____

c) Soup strainer: _____

d) Fruit/nut cracker: _____

e) All-purpose: _____

f) Knife/hook: _____

29. Write one example for each foot adaptation listed below. The first one is done for you.

a) Perching: _____robin_____

b) Walking: _____

c) Climbing: _____

d) Wading: _____

e) Swimming: _____

f) Tearing: _____

ANIMAL TAXONOMY FINAL TEST (continued)

30. What are the four major characteristics of mammals?

    a) _____

    b) _____

    c) _____

    d) _____

31. Give one example for each subclass of mammal listed below.

    a) Monotreme: _____

    b) Marsupial: _____

    c) Placental: _____

32. Name any four of the six orders of placental mammals that were presented in class (there are 17 orders all together).

    a) _____

    b) _____

    c) _____

    d) _____

33. Primates are divided into two major groupings. What are they?

    a) _____

    b) _____

34. Of these two groupings, in which one do monkeys and anthropoid apes

    belong? _____

35. Name two kinds of anthropoid apes.

    a) _____

    b) _____

36. In class you were given seven facts about mammals, other than the four major characteristics. Write down any two of these seven facts.

    a) _____

    b) _____

37. What is the most highly developed creature on earth?

    _____

ANSWERS TO ANIMAL TAXONOMY FINAL TEST

1. _5_ order _8_ species _1_ kingdom _4_ class
   _3_ subphylum _7_ genus _6_ family _2_ phylum

2. a) body covering: moist, slimy scales
   b) heating system: ectothermic (cold-blooded)
   c) respiration system: gills only
   d) reproduction system: eggs laid in water

3. fish with jaws

4. a) bony fishes
   (1) higher bony fishes
   (2) lungfish
   b) cartilaginous fishes
   (1) rays
   (2) sharks

5. primitive, two-chambered heart

6. a fish's "nostrils"; organ for smelling

7. swim bladder

8. lateral line

9. a) body covering: moist smooth skin
   b) heating system: ectothermic (cold-blooded)
   c) respiration system: gills when young/lungs when adult
   d) reproduction system: jelly-like eggs, laid in water

10. a) salamanders and newts
    b) frogs and toads
    c) caecilians

11. during the transition from gill-breather to lung-breather

12. to avoid dehydration; eggs do not have shells

13. a) body covering: dry scales
    b) heating system: ectothermic (cold-blooded)
    c) respiration system: lungs only
    d) reproduction system: leathery-shelled eggs laid on land

14. a) snakes and lizards (squamata)
    b) turtles (testudines)
    c) crocodilians (loricata)
    d) tuatara

15. yes

16. to "feel" and "taste" the air; to sense vibrations and heat

17. no

18. a) carapace
    b) plastron

19. a) lizards have eyelids and snakes do not
    b) lizards have ears and snakes do not

20. a) body covering: feathers
    b) heating system: endothermic (warm-blooded)
    c) respiration system: lungs only
    d) reproduction system: brittle-shelled eggs

21. 27

22. See the list of birds for each category in the taxonomy outline.

23. a) contour
    b) flight
    c) down

24. molt

25. to oil feathers for waterproofing

26. four-chambered heart

27. a) no teeth
    b) feathers instead of fur
    c) hollow bones
    d) very small stomach

28. (These are examples; there are many correct answers.)
    a) cardinal
    b) hummingbird
    c) flamingo
    d) parrot
    e) crow
    f) hawk

29. (These are examples; there are many correct answers.)
    a) robin
    b) ostrich
    c) woodpecker
    d) crane
    e) duck
    f) eagle

30. a) body covering: hair or fur at some time in life
    b) heating system: endothermic (warm-blooded)
    c) respiration system: lungs only
    d) reproduction system: live birth (except the two monotremes)

31. a) platypus or spiny anteater
    b) koala, kangaroo, opossum (among others)
    c) almost any mammal

32. a) rodentia
    b) artiodactyla
    c) perissodactyla
    d) cetacea
    e) carnivora
    f) primate

33. a) prosimians
    b) higher primates

34. higher primates

35. a) gorilla
    b) chimpanzee
    c) orangutan
    d) gibbon

36. See the list of facts about mammals in the taxonomy outline.

37. human beings (homo sapiens)

ANIMAL RESEARCH PROJECTS
Teacher Preview

Project Topics: Vertebrate Animals
Entomology
Oceanography

General Explanation: These three research projects are virtually identical in structure, with each offering a different subject area to study while requiring the use of a basic set of research and presentation skills. It is suggested that only one, or two at most, be used with a group of students in any given school year, to avoid repetition and potential boredom.

Length of Project: 17 hours

Level of Independence: Intermediate

Goals:

1. To allow students to learn about any of a wide variety of animals.
2. To require the use of research skills as students learn about specific animals they have chosen.
3. To promote the concept of "kids teaching kids."
4. To place emphasis on independent learning.

During This Project Students Will:

1. Select one specific animal to study as a research topic.
2. Combine the research skills they have mastered and apply them to individualized projects.
3. Follow an outline to complete the project requirements.
4. Make final presentations to the class.
5. Evaluate their own work on a "Self-Evaluation Sheet."

Skills:

Preparing bibliographies	Sentences
Collecting data	Controlling behavior
Library skills	Following project outlines
Listening	Individualized study habits
Making notecards	Persistence
Summarizing	Sharing space

Neatness
Spelling
Writing
Organizing
Setting objectives
Selecting topics
Divergent-convergent-evaluative thinking
Following and changing plans
Identifying problems
Meeting deadlines
Working with limited resources
Accepting responsibility
Grammar
Paragraphs

Taking care of materials
Time management
Personal motivation
Self-awareness
Sense of "quality"
Setting personal goals
Creative expression
Creating presentation strategies
Drawing/sketching/graphing
Poster making
Public speaking
Self-confidence
Teaching others
Handwriting
Concentration

Handouts Provided:

- "Student Introduction to the Animal Research Project"
- "Assignment Sheet" for each area of study
- "Self-Evaluation Sheet" for each area of study
- Teacher's Introduction to the Student Research Guide (optional; see Appendix)
    a. "Notecard Evaluation"
    b. "Poster Evaluation"
    c. "Oral Presentation Evaluation"
- Student Research Guide (optional; see Appendix)
    a. "Bibliographies"
    b. "Notecards and Bibliographies"
    c. "The Card Catalog"
    d. *"Readers' Guide to Periodical Literature"*
    e. "Audio-Visual and Written Information Guides"
    f. "Where to Go or Write for Information"
    g. "Poster Display Sheet"
    h. "Things to Check Before Giving Your Presentation"
    i. "Visual Aids for the Oral Presentation"
    j. "Things to Remember When Presenting Your Project"
    k. "Daily Log"

PROJECT CALENDAR:

HOUR 1: _____	HOUR 2: _____	HOUR 3: _____
Discussion of the project; students receive their assignment sheets and "Introduction to the Animal Research Project" handouts.	Rules for working on independent projects in the classroom are explained. Students begin searching for topics.	Students begin working on their projects in class. Topic choices are turned in.
PREPARATION REQUIRED HANDOUT PROVIDED	PREPARATION REQUIRED	STUDENTS TURN IN WORK
HOUR 4: _____	HOUR 5: _____	HOUR 6: _____
Work on projects.	Work on projects.	Work on projects.
NEED SPECIAL MATERIALS	NEED SPECIAL MATERIALS	NEED SPECIAL MATERIALS
HOUR 7: _____	HOUR 8: _____	HOUR 9: _____
Work on projects.	Work on projects.	Work on projects.
NEED SPECIAL MATERIALS	NEED SPECIAL MATERIALS	NEED SPECIAL MATERIALS

PROJECT CALENDAR:

HOUR 10: _____	**HOUR 11:** _____	**HOUR 12:** _____
Work on projects.	Work on projects.	Finish projects.
NEED SPECIAL MATERIALS	NEED SPECIAL MATERIALS	NEED SPECIAL MATERIALS
HOUR 13: _____	**HOUR 14:** _____	**HOUR 15:** _____
Students make presentations to the class and receive self-evaluation forms.	Students make presentations to the class and receive self-evaluation forms.	Students make presentations to the class and receive self-evaluation forms.
STUDENTS TURN IN WORK HANDOUT PROVIDED	STUDENTS TURN IN WORK HANDOUT PROVIDED	STUDENTS TURN IN WORK HANDOUT PROVIDED
HOUR 16: _____	**HOUR 17:** _____	**HOUR 18:** _____
Students make presentations to the class and receive self-evaluation forms.	Students turn in self-evaluation forms. Discussion about the skills used in this project: why they are important and how they could be used to study other things.	
STUDENTS TURN IN WORK HANDOUT PROVIDED	STUDENTS TURN IN WORK	

Lesson Plans and Notes

HOUR 1: Give students the assignment sheet and the "Student Introduction to the Animal Research Project" handout; discuss these point-by-point. Explain the grading system that will be used for this project and examine the evaluation forms. It is a good idea to present examples of each of the project requirements: notecards, bibliography cards, a poster, a diorama, a written report, and any other visual aids that may be effective in describing the project. It is helpful to have an overhead transparency of the student assignment sheet for use during this hour; an overhead transparency of notecards and bibliographies is also convenient. Finally, have a collection of typical reference books and various animal books available to show the class. Read some excerpts to show what kinds of information might be found in each type of source.

HOUR 2: If this is the first individualized project for these students, establish the ground rules for working independently in the classroom. Students must have access to plenty of information about animals: either provide a mini-library in the room, make arrangements for students to make regular trips to the school library, or plan a major research trip to a local public library. Spend Hour 2 discussing the materials that are available and looking at some of the books. Students must choose a topic by the end of the next hour.

HOUR 3: Students may begin working on their projects in class, or spend the hour deciding on final topics. Topic choices are turned in at the end of the hour.

Note:

- Be sure to evaluate the topic choices that students make. Don't let anyone pursue a topic that has insufficient information available. It's also a good idea to limit the number of people who may study any one topic.

HOURS 4-12: Students continue to work on their projects independently. Carefully monitor their progress to ensure that the proper amount of time has been provided. Additional requirements may be made during these hours if you want: (1) a rough draft report, (2) an outline of the report, (3) notecards handed in before the visual display is started, (4) a rough sketch of the poster or diorama before it is begun, (5) the finished poster and report handed in and graded before the oral report is given, and so forth. Have materials available during these hours for making visual displays: posterboard, markers, paints, rulers, whatever can be provided.

HOURS 13-16: Students make presentations to the class about their research while you grade reports from the back of the room as they are given. Students receive self-evaluation forms after they complete their reports but they do not hand them in until everyone has given a presentation.

HOUR 17: Students turn in their self-evaluation forms. This final hour gives you an opportunity to help students see the ultimate value of the project they have just completed. Discuss the skills that are necessary for conducting a research project, producing a visual display, and making an oral report. Ask students if they have a different perspective on independent learning than they had when the project began. Spend time explaining how students who have gone through the experience of a research project such as this are prepared to tackle projects that allow even more independence. Then discuss careers in science: how do scientists make use of research skills? Why is it important for people who want to learn about certain areas of science to improve their research capabilities as much as possible? Why do scientists and people who study science need to use the higher level thinking skills of knowledge, comprehension, application, analysis, synthesis, evaluation?

General Notes About These Projects:

- There are a number of teaching aids in the Student Research Guide (Appendix) that may be useful. These materials are optional and are not necessary for the successful completion of the project. However, it is worth the effort to look them over, especially the student checklists and the "Daily Log" handout, which is an excellent way for students to keep track of their own progress. There are also several evaluation forms in the Teacher's Introduction to the Student Research Guide that may be helpful.

- For students to benefit most from this project they should already be familiar with the five classes of vertebrate animals, and with such terms as "food web," "adaptation," "habitat," "reproduction," "migration," and so forth. If you have not taught the course on "Animal Taxonomy" from this book, you may want to provide the classification outline, and the "Animal Terms and Definitions" handout from the course for your students to use as a reference.

- A "Self-Evaluation Sheet" is included after the assignment sheet for each project. Each self-evaluation form is slightly different from the others; they vary in point value from 100 to 200. You can use them in a variety of ways, from gauging students' impressions of their own work to designating a percent of the final grade. Self-evaluation forms are interesting conversation pieces for parent conference, too.

Name _____ Date _____

STUDENT INTRODUCTION TO
THE ANIMAL RESEARCH PROJECT

Learning about something on your own requires a great deal of independence and self-motivation. Choosing a topic to study, finding information about it, and then presenting what you've learned to others involves many skills. You are being given the responsibility of doing a research project because you already possess many *basic* research skills and because the experience of learning on your own teaches many skills you cannot learn in a lecture.

Keep in mind that the topic you will be studying is probably not something you will need to know about to be successful in life, but the skills used to complete this project will help you a great deal in future courses and possibly in your career as well. You will gain experience in the following areas:

1. Finding sources of information about a topic you choose to study.
2. Making notecards and bibliography cards while doing research.
3. Organizing the information from notecards into a written report.
4. Creating an interesting way to teach the rest of the class about your topic.
5. Designing a poster or other visual display that is eye-catching, informative, and complementary to your oral presentation.
6. Working from an outline to complete the requirements.
7. Meeting deadlines on time by disciplining yourself to do work, and not waiting for someone else to spoon-feed you the instructions.
8. Managing your study time so that it doesn't get used up talking with friends about the latest movie, bothering others who are trying to work, or hesitating to start because you're not sure how to go about it.
9. Solving problems as they arise and changing plans if you need to.

The specific requirements of your research project are outlined on a separate handout so that you can see the entire plan and organize your work accordingly. Once you are able to complete an assignment like this on your own it is logical that you could choose almost any topic and go about studying it. All subjects cannot be taught in school but this should not keep you from learning about those which are of special interest to you. Apply the skills presented in this project and learn on your own.

Name _____ Date _____

VERTEBRATE ANIMALS
Student Assignment Sheet

The following outline is a guide for the animal research project. Use this outline to design and create a project from start to finish. Keep in mind that you will choose a topic, do research, and organize this information into its final form, all on your own. Of course, some advice or problem-solving help may be needed along the way, but you are being graded basically on your ability to follow this outline to complete project requirements. Read this handout carefully so that you know exactly what you are expected to do.

I. Choose a topic.

    A. Your topic can be *any* vertebrate animal.

        1. It may *not* be an animal you have already studied this year.

        2. It can be an animal someone else is studying, as long as each person works alone.

II. Find information.

    A. Locate at least five sources of information and make a bibliography card for each one.

        1. Encyclopedias may be used for two sources, but no more than two.

        2. Other sources of information: books, magazines, other periodicals, pamphlets, letters, experts, organizations, zoos, nature centers, veterinarians, television.

    B. Use these five sources to find twenty-five facts about your animal. Specifically, you need to:

        1. Identify some of the animal's adaptations.

        2. Find out how this animal fits into a food web. This means that you describe what it eats, what eats it, and how all these animals and plants fit together in a natural balance.

        3. Describe the animal's natural habitat.

        4. Find information about how the animal reproduces: mating habits, type of birth, special territory, care of young, and anything else you can find.

        5. Locate what you can about life span, social behavior, migration, unusual habits, territory needs, population, where the animal lives in the world, and anything else of interest.

    Record each fact on a *separate* notecard that has been properly headed.

    C. You do not have to take five facts from each source; some sources may give only one or two facts and others may give ten or more.

    D. Each fact should be simple enough to explain or describe on one notecard. At times information will have to be summarized in a few simple statements.

    E. Thoroughly understand each of your facts. Don't just copy information out of a book; if you don't understand, find out!

III. Design a project presentation.

    A. The presentation can take many forms but it must be based on research and teach about the five areas described under point II.B of this handout.

    B. You will present a report orally to the class *and* produce some type of visual display to go with the presentation. This display can take many forms:

        1. A mobile display.

        2. A slide show.

        3. A poster display.

        4. A diorama showing the animal in its natural habitat.

        5. A display of specimens.

        6. A guest speaker (you must have an active part).

        7. A display of charts and graphs.

        8. A picture, drawing, or photographic display.

        9. Don't forget that the most creative approach is to combine two or more of the above methods. You may come up with other ideas of your own.

IV. Complete final project requirements.

    A. Give your presentation.

    B. Hand in

        1. Five properly made bibliography cards.

        2. Twenty-five well-written facts, each on a separate notecard with a reference to the bibliography card that tells the source of the information.

        3. The visual display that was used with the oral presentation.

    C. Fill out a self-evaluation form which rates how well you think you did on this project. Fill this out after your presentation and hand it in after *everyone* has given a report. This completes the research project.

VERTEBRATE ANIMALS
Self-Evaluation Sheet

I. Classroom Work

A. Evaluate your classroom attitude and behavior. (Did you take the project seriously and follow the rules?) _____ (0–10 pts.)

B. Evaluate how well you took care of materials and helped keep the room clean. _____ (0–5 pts.)

C. Evaluate your behavior during other students' presentations. _____ (0–5 pts.)

TOTAL POINTS _____ (0–20 pts.)

II. Following the Outline

A. Evaluate your notecards for neatness, accuracy, and whether they were properly made (you must have at least twenty-five). _____ (0–5 pts.)

B. Evaluate your bibliography cards for neatness, accuracy, and whether they were properly made (you must have at least five). _____ (0–5 pts.)

C. Evaluate how well you researched and recorded information about your animal's:
1. adaptations _____ (0–5 pts.)
2. food web _____ (0–5 pts.)
3. habitat _____ (0–5 pts.)
4. reproduction _____ (0–5 pts.)
5. life span, social behavior, migration, unusual habits, territory needs, population, and so forth _____ (0–5 pts.)

D. Rate yourself on how "intelligent" you were about choosing a good topic. _____ (0–5 pts.)

E. Rate yourself on how much effort you put into your animal study project. _____ (0–10 pts.)

TOTAL POINTS _____ (0–50 pts.)

Name _____ Date _____

VERTEBRATE ANIMALS
Self-Evaluation Sheet (continued)

III. Presentation

    A. Evaluate your eye contact, voice projection, and speaking
style.   _____ (0–5 pts.)

    B. Evaluate how well you used your visual aids (posters,
dioramas, displays).   _____ (0–5 pts.)

    C. Evaluate how well you *taught* the rest of the class about your
animal's adaptations, food web, habitat, reproduction, and
other facts.   _____ (0–5 pts.)

    D. Evaluate the neatness, spelling, and accuracy of information
on your visual aid.   _____ (0–5 pts.)

    E. How well did you handle questions from the class at the
conclusion of your presentation?   _____ (0–5 pts.)

    F. Do you feel that you were prepared to give this presentation?
Did it go smoothly?   _____ (0–5 pts.)

TOTAL POINTS _____ (0–30 pts.)

FINAL SCORE _____ (100 points possible)

Name _____ Date _____

ENTOMOLOGY
Student Assignment Sheet

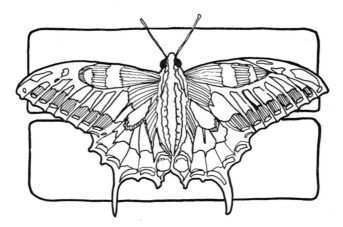

The following outline is a guide for the insect research project. Use this outline to design and create a project from start to finish. Keep in mind that you will choose a topic, do research, and organize this information into its final form, all on your own. Of course, some advice or problem-solving help may be needed along the way, but you are being graded basically on your ability to follow this outline to complete project requirements. Read this handout carefully so that you know exactly what you are expected to do.

I. Choose a topic.

A. You will study some type of insect.

B. You can choose any insect, but be sure there is enough information available before you make a final decision.

II. Find information.

A. Locate at least five sources of information and make a bibliography card for each one.

1. Encyclopedias may be used for two sources, but no more than two.

2. Other sources of information: books, magazines, other periodicals, pamphlets, letters, nature centers, organizations, museums, entomologists, hobbyists, zoos, university or college zoology departments.

B. Use these five sources to find twenty-five facts about your animal. Specifically, you need to:

1. Find out how your insect fits into a food web. This means that you describe what the insect eats, what eats it, and how all these animals and plants fit together in a natural balance.

2. Describe the insect's natural habitat.

3. Find information about how the insect reproduces: mating habits and seasons, special environments needed, and other interesting facts about it.

4. Describe the life cycle of the insect (this is connected with reproduction).

5. Describe the insect's body parts.

6. Identify some of the insect's adaptations.

Record each fact on a *separate* notecard that has been properly headed.

C. You do not have to take five facts from each source; some sources may give only one or two facts and others may give ten or more.

D. Each fact should be simple enough to explain or describe on one notecard. At times information will have to be summarized in a few simple statements.

E. Thoroughly understand each of your facts. Don't just copy information out of a book; if you don't understand, find out!

III. Design a project presentation.

A. The presentation can take many forms but it must be based on research about the six areas described under point II.B of this handout.

B. You will present a report orally to the class *and* produce some type of visual display to go with the presentation. This display can take many forms:

1. A mobile display.

2. A slide show.

3. A picture collection.

4. A poster display.

5. Charts and graphs.

6. A diorama showing the insect in its natural habitat.

7. A demonstration: for example, you may produce a puppet show, design a display of specimens, explain something like "metamorphosis" with drawn illustrations, or build a model that shows body parts.

8. Don't forget that the most creative approach is to combine two or more of the above methods. If you have other ideas you may use them also.

IV. Complete final project requirements.

A. Give your presentation.

B. Hand in

1. Five properly made bibliography cards.

2. Twenty-five well-written facts, each on a separate notecard with a reference to the bibliography card that tells the source of the information.

3. The visual display that was used with the oral presentation.

C. Fill out a self-evaluation form which rates how well you think you did on this project. Fill this out after your presentation and hand it in after *everyone* has given a report. This completes the research project.

ENTOMOLOGY
Self-Evaluation Sheet

I. Classroom Work
    A. Evaluate your classroom attitude. _____ (0–5 pts.)
    B. Evaluate your classroom behavior. _____ (0–5 pts.)
    C. Evaluate how well you took care of materials and helped keep the
        room clean. _____ (0–5 pts.)
    D. Evaluate your behavior during other students' presentations. _____ (0–5 pts.)
                      TOTAL POINTS _____ (0–20 pts.)

II. Following the Outline
    A. Are your twenty-five notecards done properly? _____ (0–5 pts.)
    B. Are your five bibliography cards done properly? _____ (0–5 pts.)
    C. Evaluate how well you researched and recorded information about
       your insect's:
       1. food web _____ (0–5 pts.)
       2. habitat _____ (0–5 pts.)
       3. adaptations _____ (0–5 pts.)
       4. reproduction _____ (0–5 pts.)
       5. life cycle _____ (0–5 pts.)
       6. body parts _____ (0–5 pts.)
    D. Rate yourself on how "intelligent" you were about topic selection. _____ (0–5 pts.)
    E. Rate yourself on how much effort you put into your entomology
       project. _____ (0–5 pts.)
                      TOTAL POINTS _____ (0–50 pts.)

III. Oral Presentation
    A. Evaluate your eye contact. _____ (0–5 pts.)
    B. Evaluate your voice projection. _____ (0–5 pts.)
    C. Evaluate how well you used your visual aids. _____ (0–5 pts.)
    D. How well did you present information about the insect's:
       1. food web _____ (0–5 pts.)
       2. habitat _____ (0–5 pts.)
       3. adaptations _____ (0–5 pts.)
       4. reproduction _____ (0–5 pts.)
       5. life cycle _____ (0–5 pts.)
       6. body parts _____ (0–5 pts.)
    E. Evaluate the accuracy of the information in your presentation. _____ (0–5 pts.)
    F. Did you present the twenty-five facts from your notecards? _____ (0–5 pts.)
    G. Do you feel that you were prepared to give this presentation? Did
       it go smoothly? _____ (0–5 pts.)
    H. Did you choose a topic that could be researched well enough to
       produce a quality presentation? _____ (0–5 pts.)
                      TOTAL POINTS _____ (0–65 pts.)

ENTOMOLOGY
Self-Evaluation Sheet (continued)

IV. Visual Presentation (posters, dioramas, displays)
- A. Evaluate the neatness of your visual presentation.   _____ (0–5 pts.)
- B. Evaluate your spelling.   _____ (0–5 pts.)
- C. Evaluate the accuracy of the information in your visual presentation.   _____ (0–5 pts.)
- D. How well does your visual presentation present information about the insect's:
  - 1. food web   _____ (0–5 pts.)
  - 2. habitat   _____ (0–5 pts.)
  - 3. adaptations   _____ (0–5 pts.)
  - 4. reproduction   _____ (0–5 pts.)
  - 5. life cycle   _____ (0–5 pts.)
  - 6. body parts   _____ (0–5 pts.)
- E. Evaluate how "interesting" your visual presentation is to look at and study.   _____ (0–5 pts.)
- F. Does your visual presentation include twenty-five facts from your notecards?   _____ (0–5 pts.)
- G. How much effort did you put into your visual display?   _____ (0–5 pts.)
- H. Did you choose a topic that could be researched well enough to produce a quality visual presentation?   _____ (0–5 pts.)

TOTAL POINTS _____ (0–65 pts.)

FINAL SCORE _____ (200 points possible)

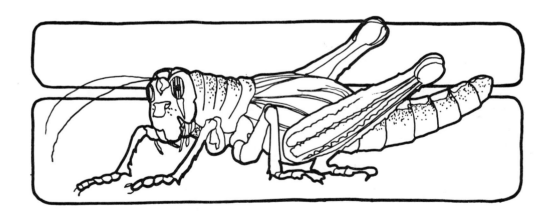

Name _____ Date _____

OCEANOGRAPHY
Student Assignment Sheet

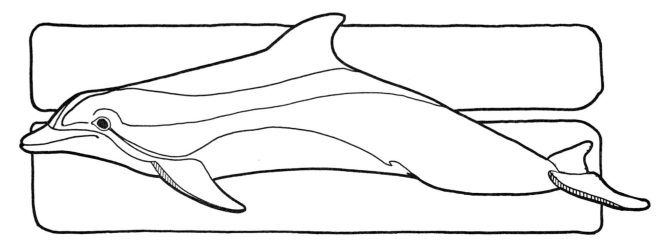

The following outline is a guide for the oceanography project. Use this outline to design and create a project from start to finish. Keep in mind that you will choose a topic, do research, and organize this information into its final form, all on your own. Of course, some advice or problem-solving help may be needed along the way, but you are being graded basically on your ability to follow this outline to complete project requirements. Read this handout carefully so that you know exactly what you are expected to do.

I. Choose a topic.

   A. Your topic can be any animal that lives in the ocean.
   B. Narrow the topic down to a specific species of animal, if possible. For example, you probably would not choose whales as a topic because it is too general. Instead you could choose the blue whale, the humpback whale, or possibly the killer whale. You may have to be more general for some topics if there is not a lot of information available about any one species. Use your best judgment when choosing a topic.

II. Find information.

   A. Locate at *least* five sources of information and make a bibliography card for each one.
      1. Encyclopedias may be used for two sources, but no more than two.
      2. Other sources of information: books, periodicals, films, television, pamphlets, marine research centers, aquariums, zoologists, and organizations.

   B. Use these five sources to find twenty-five facts about the animal. Specifically, you need to
      1. Identify and describe some of the animal's adaptations. This means you describe what it has that allows it to live where it does, in the way it does.
      2. Find out how this animal fits into a food web. In other words, find out what the animal eats, what eats it, and how all these animals and plants fit together in a natural balance.
      3. Describe the animal's habitat.
      4. Find information about the animal's reproduction: mating habits, type of development, care of young, and anything else you can find.

5. Locate what you can about life span, social behavior, migration, unusual habits, territory needs, population, where the animal lives in the ocean, and anything else of interest.

Record each fact on a *separate* notecard that has been properly headed.

C. You do not have to take five facts from each source; some sources may give only one or two facts and others may give ten or more.

D. Each fact should be simple enough to explain or describe on one notecard. At times information will have to be summarized in a few simple statements.

E. Thoroughly understand each of your facts. Don't just copy information out of a book; if you don't understand, find out!

III. Design a project presentation.

A. The presentation can take many forms, but it must be based on research and teach about the five areas described under point II.B of this handout.

B. You will present a report orally to the class *and* produce some type of visual display to go with the presentation. This display can take many forms:

1. A mobile display.

2. A slide show.

3. A picture collection.

4. A poster display.

5. Charts and graphs.

6. A diorama, showing the animal in its natural habitat.

7. A demonstration: for example, you may produce a puppet show, design a display of specimens, explain something like "adaptations" with drawn illustrations or build a model that shows how the animal is equipped to live in the ocean.

8. Don't forget that the most creative approach is to combine two or more of the above methods. If you have other ideas you may use them also.

IV. Complete final project requirements.

A. Give your presentation.

B. Hand in

1. Five properly made bibliography cards.

2. Twenty-five well-written facts, each on a separate notecard with a reference to the bibliography card that tells the source of the information.

3. The visual display that was used with the oral presentation.

C. Fill out a self-evaluation form which rates how well you think you did on this project. Fill this out after your presentation and hand it in after *everyone* has given a report. This completes the research project.

Name _____ Date _____

OCEANOGRAPHY
Self-Evaluation Sheet

I. Classroom Work
   A. Evaluate your classroom attitude and behavior. _____ (0–5 pts.)
   B. Evaluate how well you took care of materials and helped keep the room clean. _____ (0–5 pts.)
   C. Evaluate your behavior during other students' presentations. _____ (0–5 pts.)

   TOTAL POINTS _____ (0–15 pts.)

II. Following the Outline
   A. Evaluate your twenty-five notecards: are they done properly? _____ (0–5 pts.)
   B. Evaluate your five bibliography cards: are they done properly? _____ (0–5 pts.)
   C. Evaluate how well you researched and recorded information about the animal's *food web*. _____ (0–5 pts.)
   D. Evaluate how well you researched and recorded information about the animal's *habitat*. _____ (0–5 pts.)
   E. Evaluate how well you researched and recorded information about the animal's *adaptations*. _____ (0–5 pts.)
   F. Rate yourself on how "intelligent" you were about topic selection. _____ (0–5 pts.)
   G. Rate yourself on how much effort you put into the oceanography project. _____ (0–5 pts.)

   TOTAL POINTS _____ 0–35 pts.)

III. Oral Presentation
   A. Evaluate your eye contact. _____ (0–5 pts.)
   B. Evaluate your voice projection. _____ (0–5 pts.)
   C. Evaluate how well you used your visual aids. _____ (0–5 pts.)
   D. How well did you present information about the animal's *food web*? _____ (0–5 pts.)
   E. How well did you present information about the animal's *habitat*? _____ (0–5 pts.)
   F. How well did you present information about the animal's *adaptations*? _____ (0–5 pts.)
   G. Evaluate the accuracy of the information in your presentation. _____ (0–5 pts.)
   H. Did you present the twenty-five facts from your notecards? _____ (0–5 pts.)
   I. Do you feel that you were prepared to give this presentation? Did it go smoothly? _____ (0–5 pts.)
   J. Did you choose a topic that could be researched well enough to produce a quality presentation? _____ (0–5 pts.)

   TOTAL POINTS _____ (0–50 pts.)

IV. Visual Presentation (posters, dioramas, displays)
    A. Evaluate the neatness of your visual display.   _____ (0–5 pts.)
    B. Evaluate your spelling.   _____ (0–5 pts.)
    C. Evaluate the accuracy of the information in your visual display.   _____ (0–5 pts.)
    D. How well does your visual display present information about the animal's *food web?*   _____ (0–5 pts.)
    E. How well does your visual display present information about the animal's *habitat?*   _____ (0–5 pts.)
    F. How well does your visual display present information about the animal's *adaptations?*   _____ (0–5 pts.)
    G. Evaluate how "interesting" your visual display is to look at and study.   _____ (0–5 pts.)
    H. Does your visual display include twenty-five facts from your notecards?   _____ (0–5 pts.)
    I. How much effort did you put into your visual display?   _____ (0–5 pts.)
    J. Did you choose a topic that could be researched well enough to produce a quality visual display?   _____ (0–5 pts.)

TOTAL POINTS   _____ (0–50 pts.)

FINAL SCORE   _____ (150 points possible)

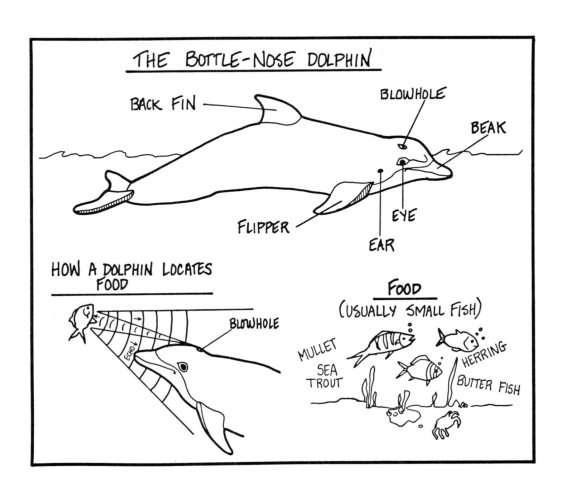

SHOEBOX DIORAMAS

Teacher Preview

Project Topics: Animals in the Wild
Imaginary Animals

General Explanation: There are two diorama projects described here. Because the first part of the Teacher Preview (length of project, goals, objectives, skills, handouts provided, and general notes) is the same for both projects, this material is presented once to conserve space and avoid repetition. The project calendar and lesson plans and notes for "Animals in the Wild" come next, followed by the same material for "Imaginary Animals." Finally, the assignment sheets and evaluation sheet for the projects are provided.

Length of Project: 15 hours

Level of Independence: Intermediate

Goals:

1. To help students learn about animals in their natural environments, and the adaptations that allow them to survive.
2. To require the use of research skills as students learn about specific animals ("Animals in the Wild" only).
3. To emphasize information about animal classification ("Imaginary Animals" only).
4. To promote the concept of "kids teaching kids."
5. To place emphasis on independent learning.
6. To combine artistic endeavors with academic objectives.

During This Project Students Will:

1. Select animals and conduct research ("Animals in the Wild" only).
2. Create imaginary animals equipped with appropriate adaptations to survive in a specified environment ("Imaginary Animals" only).
3. Combine the research skills they have mastered and apply them to individualized projects.
4. Follow an outline to complete the project requirements.
5. Construct "shoebox dioramas."

Skills:

Preparing bibliographies

Collecting data

Library skills

Listening

Summarizing

Writing

Organizing

Outlining

Setting objectives

Divergent-convergent-evaluative thinking

Following and changing plans

Identifying problems

Meeting deadlines

Working with limited resources

Accepting responsibility

Concentration

Controlling behavior

Neatness

Spelling

Following project outlines

Individualized study habits

Persistence

Sharing space

Taking care of materials

Time management

Personal motivation

Self-awareness

Setting personal goals

Creative expression

Creating presentation strategies

Diorama and model building

Drawing/sketching/graphing

Self-confidence

Teaching others

Selecting topics

Sense of "quality"

Grammar

Handwriting

Paragraphs

Sentences

Poster making

Handouts Provided:

- "Student Assignment Sheet" for both areas of study
- "Shoebox Diorama Evaluation" (for use with either project)
- Student Research Guide (optional; see Appendix)
  — Choose the handouts that fit your needs.

General Notes About These Projects:

- Shoebox diorama projects are designed primarily to provide an opportunity to evaluate students on their ability to work independently in the classroom. They have very little to do with teaching children how to work with pipe cleaners and paint. Whether or not a student can create an accurate representation of a raccoon in a tree is of little long-term importance. However, the ability to conceive an idea, devise a plan, and create a final product *is* important, and that is what shoebox dioramas are all about.

- Artistic and creative abilities are important, though, and so shoebox dioramas also serve as a means of combining art with the academic process of research. Once students have learned factual information about animals, they are allowed to express or exhibit what they have learned.

- Shoebox dioramas are potentially very messy and can create a rather chaotic classroom atmosphere if you are not prepared with adequate materials and a firm set of rules for behavior. Students should understand the consequences of misbehaving or avoiding work.

- If you have never conducted a classroom project like this, your first effort will probably be less than perfect. Understanding this from the outset will help you learn from these experiences and be better prepared for the next time. You will be surprised at how much easier your second attempt seems after having done the project once.

- Have the following supplies on hand:

modeling clay	glue
plaster of paris	markers (fine and wide)
papier-mâché	toothpicks
construction paper	small dowel rods
poster paints	aluminum foil
watercolors	plastic wrap
tape	baby food jars
pipe cleaners	paintbrushes
clothespins	scissors
paper cups	running water
tongue depressors/popsicle sticks	utility sink
cotton balls	rags
sand and gravel	newspapers

- A simple system for checking materials out and back in should be established before the project begins.

Some Suggestions for Building Shoebox Dioramas:

1. Encourage students to make their animals as paper or cardboard cutouts that can stand upright. If "people" are to be included, they can be made from cardboard cutouts or clothespins. Don't use clay unless you have a kiln, and even then be careful and try to discourage it. Too many clay figures fall apart! There are clay substitutes on the market that work much better.

2. Line the box with foil or plastic wrap before using damp clay, papier-mâché, plaster of paris, sand or gravel.

3. Use a trigger sprayer (like a houseplant sprayer) filled with a 50% white glue/50% water solution to spray over areas covered with sand, gravel, loose grass, or any other granular or powdered substance, to hold it in place. The sprayer must be cleaned *thoroughly* after each use.

4. Put poster paints in baby food jars to be distributed around the room. Assign one or two students to help with this job.

5. Aluminum foil painted with blue watercolor makes nice water, and a small mirror is perfect as a pond or lake with papier-mâché or plaster of paris banks.

6. Shoeboxes can be viewed from the side, end, or top. They can be placed horizontally or vertically, and they can be set inside their tops to provide more foreground space. They can also be covered with their tops and have a hole, slot or window cut somewhere for viewing. Some students may want to install a light. (Be sure it isn't so hot that it creates a fire hazard. Christmas tree bulbs work well.)

7. Be sure to have *plenty* of newspapers available to cover desks and the floor.

PROJECT CALENDAR: ANIMALS IN THE WILD

HOUR 1: _____	HOUR 2: _____	HOUR 3: _____
Introduction to the project. Students are told to choose at least three animals to discuss as topics next hour. HANDOUT PROVIDED	Discussion about topic choices. Each student chooses one animal to study.	Students begin conducting research about the animal they have chosen.
HOUR 4: _____	**HOUR 5:** _____	**HOUR 6:** _____
Research continues.	Research continues; papers are due next hour.	Papers are handed in and students spend the rest of the hour planning their dioramas. Drawings of dioramas and materials lists are turned in at the end of the hour. STUDENTS TURN IN WORK
HOUR 7: _____	**HOUR 8:** _____	**HOUR 9:** _____
Students begin working on their dioramas. NEED SPECIAL MATERIALS RETURN STUDENT WORK	Work on dioramas. NEED SPECIAL MATERIALS	Work on dioramas. NEED SPECIAL MATERIALS

PROJECT CALENDAR: ANIMALS IN THE WILD

HOUR 10: _____	HOUR 11: _____	HOUR 12: _____
Work on dioramas.	Work on dioramas.	Students finish their dioramas by the end of this hour.
NEED SPECIAL MATERIALS	NEED SPECIAL MATERIALS	NEED SPECIAL MATERIALS
HOUR 13: _____	**HOUR 14:** _____	**HOUR 15:** _____
Students present their dioramas to the class.	Students present their dioramas to the class.	Discussion of the diorama project and what students learned from it. Dioramas are returned and students receive evaluations for their shoebox presentations. An evaluation sheet is provided.
STUDENTS TURN IN WORK	STUDENTS TURN IN WORK	HANDOUT PROVIDED RETURN STUDENT WORK
HOUR 16: _____	**HOUR 17:** _____	**HOUR 18:** _____

Lesson Plans and Notes
for Animals in the Wild

HOUR 1: Introduce students to the requirements of the project; distribute and explain the student assignment sheet. At the end of the hour students are told to come to class the next hour with at least three animals in mind that they would like to study (three per person).

HOUR 2: Students tell what animals they have chosen: use your best judgment to advise them whether or not their choices have adequate information available. By the end of the hour each student chooses *one* animal to study. When you have confirmed these choices, students can begin their independent research.

Note:

- At this point students should have access to reference material, either in the classroom or in the library. If possible, the following three hours should be conducted in a library, to provide access to the greatest amount of information.

HOURS 3-5: Students continue doing research and then write this information into a report to be handed in Hour 6.

HOUR 6: At the beginning of the hour students hand in their research paper (two to three pages each). They spend the rest of the hour making drawings (or, more accurately, plans) of proposed dioramas, and lists of the materials needed to build them. These drawings and lists are handed in at the end of the hour. If students have difficulty finishing by the end of the hour, the drawings and lists can be homework assignments.

HOUR 7: Return the graded research papers, drawings, and materials lists to students, and hand out the shoeboxes. (See *Note* below.) Present the classroom rules, materials checkout procedures, and cleanup instructions. Students then begin working on dioramas.

Note:

- Be sure to tell students well in advance of this hour that they will need to bring in shoeboxes to work with. Since these do take up quite a bit of room, make arrangements to store them between class hours. Try to have a few extra boxes on hand for students who are unable to bring their own.

HOURS 8-12: Students continue to work on their dioramas, which must be completed by the end of Hour 12.

HOURS 13-14: Students present their work to the class and then hand in their shoeboxes to be graded. Use the "Shoebox Diorama Evaluation Sheet" to grade students for their work on this project. It includes sections for evaluating classroom attitudes and behavior, the finished product, neatness and creativity, and the written report. Since it may take more than one night to fill these out, plan to have Hour 15 as much as a week after Hour 14.

HOUR 15: Dioramas are returned and students receive evaluations for their shoebox presentations. End the project with a discussion of what was really learned by building shoebox dioramas. Place emphasis on the importance of taking known information (knowledge) and making something new from it. Remind the students that this is what they did in planning, designing, and constructing their dioramas, based upon their knowledge of animals. Also focus the discussion on self-evaluation as a valuable skill. Who decides what is "good enough" for a project like this? To be independent learners, students must be able to judge, or assess, their own work; *they* decide when their work is good enough, based upon their own sense of quality.

PROJECT CALENDAR: IMAGINARY ANIMALS

HOUR 1: _____ Introduction to the project. A drawing is conducted to help determine what kinds of animals students will invent. HANDOUT PROVIDED PREPARATION REQUIRED	**HOUR 2:** _____ Students work on making sketches of imaginary animals.	**HOUR 3:** _____ Work on sketches.
HOUR 4: _____ Work on sketches.	**HOUR 5:** _____ Sketches (or drawings) are handed in, students begin planning their dioramas. STUDENTS TURN IN WORK	**HOUR 6:** _____ Top-view drawings (plans) of dioramas and material lists are turned in and students receive their imaginary animal drawings back. STUDENTS TURN IN WORK RETURN STUDENT WORK
HOUR 7: _____ Students receive their plans back. This is followed by a discussion of how the project is to proceed. RETURN STUDENT WORK	**HOUR 8:** _____ Students begin working on their dioramas. NEED SPECIAL MATERIALS	**HOUR 9:** _____ Work on dioramas. NEED SPECIAL MATERIALS

PROJECT CALENDAR: IMAGINARY ANIMALS

HOUR 10: _____ Work on dioramas. NEED SPECIAL MATERIALS	**HOUR 11:** _____ Work on dioramas. NEED SPECIAL MATERIALS	**HOUR 12:** _____ Students finish their dioramas by the end of this hour. NEED SPECIAL MATERIALS
HOUR 13: _____ Students present their dioramas to the class. STUDENTS TURN IN WORK	**HOUR 14:** _____ Students present their dioramas to the class. STUDENTS TURN IN WORK	**HOUR 15:** _____ Discussion of the diorama project and what students learned from it. Dioramas are returned and students receive evaluations for their shoebox presentations. An evaluation sheet is provided. HANDOUT PROVIDED RETURN STUDENT WORK
HOUR 16: _____	**HOUR 17:** _____	**HOUR 18:** _____

Lesson Plans and Notes
for Imaginary Animals

HOUR 1: Give out the assignment sheets and discuss the entire project. When students understand how the project works, conduct "drawings" to determine what "kinds" of animals they will create. Place the subfactors described on the handout in a hat (one set at a time) and have students draw to determine what climate, activity period, diet, habitat, and social grouping their imaginary animals must be equipped for. The *purpose* of this project is to teach about *adaptations* and *classification*.

Notes:

- To determine a specific set of subfactors (from the student handout) for each student, conduct five separate "drawings" out of a hat. Start with "climates." If you have thirty students, put in a hat ten slips labeled "polar," ten labeled "temperate," and ten labeled "tropic." Pass the hat and let each student pick a slip. Students should record their climate subfactor on their handouts and then draw for "activity period" (nocturnal, diurnal, or crepuscular). At the end of the fifth drawing each student will have a specific set of five subfactors that help define the animal to be invented.

- Before the project begins, explain a few things about social and solitary animals, and about territory. Solitary animals are usually predators with a fixed territory. Social animals are usually prey with no specific territory (except at mating time).

HOUR 2: Students begin working on their sketches of imaginary animals. They are encouraged to be imaginative while following the guidelines in the project outline. All of the information asked for on the outline must be included on the drawing.

Notes:

- Allow students to invent all kinds of animals, even if they are outlandish. Be critical of only one major thing: if an animal *has* something it must be *used* for something, and if an animal lives in a certain way it must be equipped for that lifestyle. In other words, place major emphasis on adaptations. If an animal is polar it must have some means of staying warm, even if it's a ridiculous way, like flying into volcanoes. The five factors from the student handout (climate, activity period, diet, habitat, social group) must be carefully considered and accounted for in each imaginary animal.

- Before creating an imaginary animal, each student must classify the animal he or she wishes to invent as one of the five *real* classes of vertebrates: fish, amphibian, reptile, bird, or mammal. See the "Animal Taxonomy" project for more information about classification and adaptations. Each imaginary animal will belong to one of the five classes and must have the four basic characteristics of that class.

HOURS 3-4: Students continue to work on their sketches.

HOUR 5: Students hand in their imaginary animal sketches (drawings) at the beginning of the hour. They spend the rest of the hour planning dioramas by working on top-view drawings and material lists.

HOUR 6: Students hand in their top-view drawings and material lists, and receive their graded imaginary animal drawings back. Spend the hour in a general discussion about the kinds of things that could have been done to improve the drawings; display some of the unique and creative ideas that were turned in.

HOUR 7: Top-view drawings and materials lists are handed back. Spend the hour discussing what is realistically possible in a project like this: emphasize simplicity because complex ideas are often difficult to bring to completion. Present the classroom rules and procedures for working on individualized projects and tell students to come to the next class prepared to work on dioramas. Shoeboxes should be brought to class *before* Hour 8 so you know each student will have one.

HOURS 8-12: Students work on their dioramas. These must be completed by the end of Hour 12.

HOURS 13-14: Students present their work to the class and then hand in their shoeboxes to be graded. Use the "Shoebox Diorama Evaluation Sheet" to grade students for their work on this project. It includes sections for evaluating classroom attitudes and behavior, the finished product, neatness and creativity, and the written report. Since it may take more than one night to fill these out, plan to have Hour 15 as much as a week after Hour 14.

HOUR 15: Dioramas are returned and students receive evaluations for their shoebox presentations. End the project with a discussion of what was really learned by building shoebox dioramas. Place emphasis on the importance of taking known information (knowledge) and making something new from it. Remind the students that this is what they did in planning, designing, and constructing their dioramas, based upon their knowledge of animals. Also focus the discussion on self-evaluation as a valuable skill. Who decides what is "good enough" for a project like this? To be independent learners, students must be able to judge, or assess, their own work; *they* decide when their work is good enough, based upon their own sense of quality.

© 1987 by The Center for Applied Research in Education, Inc.

Name _____ Date _____

ANIMALS IN THE WILD
Student Assignment Sheet

A shoebox diorama is a visual display that shows a scene or an idea in three dimensions. It is more difficult to create a shoebox diorama than it is to produce a drawing on paper because of the added dimension of depth, but this also means there are many more effects that can be created with this type of display. Imagine, for example, the scene of a bobcat leaping over a snow-covered log in pursuit of a rabbit. On paper this can be drawn and colored to look good, but in a diorama the log will be three-dimensional; cotton balls can be torn apart and used as snow throughout the scene, and a background can be painted on the inner walls of the shoebox to show trees or fields which add the feeling of depth to this display.

When you create a shoebox diorama, set your imagination free as you plan the scene and decide on materials to use. Just remember that lightweight materials work the best since the cardboard box you will be using isn't very strong.

Your assignment is to choose an animal that lives in the wild (in other words, a "nondomestic animal"), write a two- to three-page report about that animal and build a shoebox diorama showing the animal in its natural habitat. This handout is an outline which describes what you need to do to complete the project. Your grade will be based upon how well you follow the project outline and work in class.

I. Write a two- or three-page report about your animal. Use at least two sources of information and make a bibliography for each. This paper will do the following:

A. Describe three adaptations the animal has that help it live in its natural habitat.
B. Describe the animal's natural habitat:
 1. Where is its natural home on a map?
 2. Is the area cold, rainy, dry, tropical, and so forth?
 3. Does it live in a forest, desert, swamp, grassland, lake, or so on?
 4. Does it live in trees, on the ground, underground, in the water?
 5. What other animals and plants live in this habitat?

C. Describe the animal's eating habits: is the animal a carnivore, herbivore, or omnivore? What does it eat?

D. Include anything else that is of interest.

II. Make at least two drawings of what the diorama is going to look like. These drawings will:

A. Show everything you actually plan to include in your diorama with a *front-view* drawing.

B. Show what the background will look like.

C. Show what colors you plan to use (sky, ground, animals, water, mountains, sun, trees).

D. Show where each item will be located in the shoebox using a *top-view* drawing.

III. Make a list of materials you will need to build the diorama.

IV. Build a shoebox diorama showing the animal in its natural habitat. The diorama should include:

A. A background which adds to the total scene you are trying to create. In other words, it should be more than a blank piece of construction paper.

B. At least two examples of the animal.

C. A proper setting. Show the animal's natural habitat as accurately as possible.

D. Proper perspective. Show the different sizes of things, so that a lion doesn't look bigger than a buffalo, and a monkey doesn't look taller than a tree.

E. A neatly decorated exterior. The outside of the shoebox should be attractive and neat.

F. A simple design. Don't clutter the diorama with too many extra things. Try to show the animals in a simple setting. Keeping the diorama simple will allow you to put more quality into what you do.

G. A title, such as "The White-Tailed Deer."

V. Present your diorama to the class and explain what you have learned about this animal. Describe the diorama scene and what it shows.

Name _____ Date _____

IMAGINARY ANIMALS
Student Assignment Sheet

THE ARBOREAL WATSIT

"Imaginary Animals" is a fun way to learn about animal adaptations; it calls upon your imagination, creativity, and understanding of animal characteristics to produce a display that shows an animal you have invented. Imaginary animals can look like anything, be any color, any shape, and have the strangest characteristics, but all for very good reasons which you will describe in a one-page report.

Before the diorama is started, you must turn in a sketch to show what the animal will look like. Follow this outline as you work on the project. Combine your imagination and knowledge of animal adaptations to create something new.

I. Animal Characteristics and Adaptations: Before you invent an animal, five separate "drawings" will be conducted to help determine a set of characteristics and adaptations your animal will have. Each drawing will identify one specific subfactor from the list below. For example, the first drawing is for "climate." If you draw "polar," the animal you invent has to be equipped for the cold. If you come up with "nocturnal" in the "activity period" drawing, your animal must be adapted to night-time living. The remaining three drawings will identify your animal's diet, habitat, and social grouping. These things will combine to help you decide where your animal lives, what it looks like, what it eats, what it is covered with, and so forth. As you draw subfactors for your animal, record them on the last page of this handout.

A. Climate
   1. Polar (cold)
   2. Tropic (hot)
   3. Temperate (cold/hot)

B. Activity period
   1. Nocturnal (night)
   2. Diurnal (day)
   3. Crepuscular (dawn/ dusk)

C. Diet
   1. Carnivore (meat)
   2. Herbivore (plants)
   3. Omnivore (meat/ plants)

D. Habitat
   1. Aquatic (water)
   2. Terrestrial (land)
   3. Arboreal (trees)

E. Social Grouping
   1. Social (more than three in a nonfamily group)
   2. Solitary (except in mating season)

II. Project Requirements

   A. Produce a detailed sketch of an imaginary animal which includes the following:

      1. Show how the animal is adapted for each of the five subfactors you drew in Part I. In other words, create *adaptations* for your animal and label these on the sketch with brief explanations of what the adaptations are for. Label everything possible. Also give your animal an interesting name.

      2. Decide which one of the five classes of vertebrate animals this animal belongs in and explain why it fits into this class. Give it the four basic characteristics of that class.

      3. Describe the animal's environment: does it rain a lot? Is it mountainous? Are there many trees? Is there a river or lake nearby? Are there places to hide or to make dens or nests? What kinds of plants are there? What does your animal eat? Be as descriptive as possible.

      4. Invent and label at least three additional adaptations, to cover such factors as communication, reproduction, defense, locomotion, migration, camouflage, the five senses, or internal anatomy.

      5. Turn your sketch in to be checked before starting the shoebox diorama.

   B. Produce a shoebox diorama of an imaginary animal:

      1. Decide what your diorama is going to look like and make a list of all the things you want to include. For example: rocks, a stream, a hollow log, background mountains, waterfalls, a sunset, three animals, a pond, trees, tall grass, and so forth.

      2. Make a *top-view* drawing of the diorama to show where everything is going to be located. Be as detailed as possible.

      3. Make a materials list which tells everything you will need to build the display. For example: scissors; glue; blue, brown, and green construction paper; cotton balls; aluminum foil; toothpicks; papier-mâché; white posterboard; watercolors; pipe cleaners; popsicle sticks.

      4. After the top-view drawing and materials list is checked, you may begin your diorama.

      5. Include the following things in the diorama.

         a) At least three representatives of your imaginary animal: a male, a female, and a baby. Include more if you wish. These animals should be drawn on posterboard, cut out, and used as stand-up figures.

         b) The animal's natural environment should be clearly shown.

         c) Give your diorama a background (also "sidegrounds").

         d) Each of the five subfactors you picked during Part I must be clearly shown or explained:

           Climate_____

           Activity Period_____

           Diet_____

           Habitat_____

           Social Grouping_____

      6. Write a one-page report about the animal to turn in with your diorama. Tell everything you know about the animal you have invented.

      7. Present your work to the class.

© 1987 by The Center for Applied Research in Education, Inc.

SHOEBOX DIORAMA EVALUATION
Final Grade_____

I. Classroom Attitude and Behavior (30 points): This evaluation is based on how you worked in the classroom.

    A. You solved your own problems. _____ (0–5 pts.)

    B. You did not give up, even when you were stumped. _____ (0–5 pts.)

    C. You worked well around others. _____ (0–5 pts.)

    D. You worked on your own without being told. _____ (0–5 pts.)

    E. You cleaned up after yourself, especially at the end of the hour. _____ (0–5 pts.)

    F. You were patient when the teacher couldn't get to you immediately to answer a question. _____ (0–5 pts.)

             Subtotal _____ (30 points possible)

II. Finished Product (30 points): This evaluation is based on whether you worked steadily and seriously on your diorama each day, and if you turned the diorama in on time.

    A. You got to work as soon as class started each day. _____ (0–5 pts.)

    B. You followed your drawing closely. _____ (0–5 pts.)

    C. You completed the requirements of the outline. _____ (0–5 pts.)

    D. You turned your project in on time. _____ (0–10 pts.)

             Subtotal _____ (25 points possible)

III. Neatness and Creativity (30 points): This evaluation is based on your finished diorama and also on how you worked in the classroom.

    A. You used your own ideas as you constructed the diorama and did not constantly ask for help. _____ (0–5 pts.)

    B. You showed creativity in the materials you used and the objects you made. _____ (0–5 pts.)

    C. Your diorama is orderly and uncluttered. _____ (0–5 pts.)

    D. You used space effectively. In other words, you placed things in the shoebox so that they are easily seen and properly positioned. _____ (0–5 pts.)

    E. All the objects in your diorama are proportional in size. In other words, big things appear to be big and small things appear to be small. _____ (0–5 pts.)

    F. Your background and "sidegrounds" add to the total scene. _____ (0–5 pts.)

             Subtotal _____ (30 points possible)

IV. Written Report (15 points)

    A. You followed the requirements of the outline. _____ (0–5 pts.)

    B. Writing

        1. Grammar _____ (0–2 pts.)

        2. Handwriting _____ (0–2 pts.)

        3. Paragraphs _____ (0–2 pts.)

        4. Sentences _____ (0–2 pts.)

        5. Spelling _____ (0–2 pts.)

             Subtotal _____ (15 points possible)

            FINAL SCORE _____ (100 points possible)

NATURAL ENVIRONMENTS

Teacher Preview

General Explanation: This is a small-group project (four to five students per group). Each group chooses a natural environment to study. Students within the group study specific topics about the environment. Group members share information from their research to produce an informative mural. Thus a class of thirty produces six or seven murals, each depicting a different natural environment.

Length of Project: 20 hours

Level of Independence: Advanced

Goals:

1. To help students learn about some of the animals, plants, and natural environments found in their home state.
2. To have students experience "working on their own" in small groups.
3. To apply research skills.

During This Project Students Will:

1. Select natural environments to study.
2. Study topics in small groups.
3. Create a "Natural Environment Roster," which is a list of animals and plants that can be found in the environment that group members have chosen to study.
4. Choose animals and plants from the group's "Natural Environment Roster" as subjects for individual research.
5. Combine the research skills they have mastered and apply them to individualized projects.
6. Follow an outline to complete the project requirements.
7. Pool their information and produce natural environment murals that depict the animals and plants they have studied.
8. Evaluate their projects.

Skills:

Preparing bibliographies	Working with limited resources
Collecting data	Accepting responsibility
Library skills	Concentration

Listening	Controlling behavior
Making notecards	Following project outlines
Summarizing	Individualized study habits
Grammar	Persistence
Handwriting	Sharing space
Neatness	Taking care of materials
Paragraphs	Time management
Sentences	Personal motivation
Spelling	Self-awareness
Group planning	Sense of "quality"
Organizing	Setting personal goals
Setting objectives	Creative expression
Selecting topics	Creating presentation strategies
Divergent-convergent-evaluative thinking	Diorama and model building
Identifying problems	Drawing/sketching/graphing
Meeting deadlines	Poster making
Working with others	Self-confidence
Writing	Teaching others
	Following and changing plans

Handouts Provided:

- "Student Assignment Sheet"
- "Natural Environment Roster"
- "Animal and Plant Selections"
- "Natural Environment Research Guide"
- "Self-Evaluation Sheet"
- "Final Evaluation"
- Student Research Guide (optional; see Appendix)
  — Choose the handouts that meet your needs.

PROJECT CALENDAR:

HOUR 1: _____ Introduction to the project. Students receive all project handouts; small groups are determined. Grading system is explained. HANDOUTS PROVIDED	**HOUR 2:** _____ Discussion of environments. Students (in groups) hand in lists of first, second and third choices of environments to study. PREPARATION REQUIRED STUDENTS TURN IN WORK	**HOUR 3:** _____ Groups are told which environments they will be allowed to study and students begin to develop environment rosters. Students were given the "Natural Environment Roster" handout during Hour 1. NEED SPECIAL MATERIALS
HOUR 4: _____ Students continue to work on environment rosters.	**HOUR 5:** _____ Environment rosters are finished and handed in at the end of the hour. STUDENTS TURN IN WORK	**HOUR 6:** _____ Environment rosters are returned and each student chooses two topics to study individually. Completed "Animal and Plant Selections" sheets are turned in at the end of the hour. Students were given this handout during Hour 1. RETURN STUDENT WORK STUDENTS TURN IN WORK
HOUR 7: _____ Students begin individual research, based upon the "Natural Environments Research Guide" handout. Students were given this handout during Hour 1. RETURN STUDENT WORK	**HOUR 8:** _____ Research continues.	**HOUR 9:** _____ Research continues.

PROJECT CALENDAR:

HOUR 10: _____	**HOUR 11:** _____	**HOUR 12:** _____
Research continues.	Research continues. Students begin writing reports if they are ready.	Students work on written reports.
HOUR 13: _____	**HOUR 14:** _____	**HOUR 15:** _____
Students complete written reports.	Research papers are turned in to be checked and students begin planning their murals. Groups hand in rough sketches of murals at the end of the hour. STUDENTS TURN IN WORK	Research papers and rough mural sketches are returned. Students work in small groups on their murals. NEED SPECIAL MATERIALS RETURN STUDENT WORK
HOUR 16: _____	**HOUR 17:** _____	**HOUR 18:** _____
Work on murals continues. NEED SPECIAL MATERIALS	Work on murals continues. NEED SPECIAL MATERIALS	Work on murals continues. NEED SPECIAL MATERIALS

PROJECT CALENDAR:

HOUR 19: _____	HOUR 20: _____	HOUR 21: _____
Students finish their murals and complete self-evaluation forms that were handed out during Hour 1. A "Final Evaluation" form is also provided. NEED SPECIAL MATERIALS HANDOUT PROVIDED	Class discussion about independent learning, skill development, and group dynamics.	
HOUR 22: _____	**HOUR 23:** _____	**HOUR 24:** _____
HOUR 25: _____	**HOUR 26:** _____	**HOUR 27:** _____

Lesson Plans and Notes

HOUR 1: Introduce students to the entire project and its various components, and provide a time line to show when each part of the project is due. Distribute and discuss all of the handouts; explain the grading system and go over the evaluation sheets to be used for the project. The "Final Evaluation" handout should be collected at the end of the hour after the students have had an opportunity to look it over. Supervise the class as small groups are chosen, selected, appointed, assigned, or put together (it is up to you to decide how to determine who is in each group).

HOUR 2: Present students with a list of at least ten natural environments that can be found in their home state. *You are responsible for compiling and providing this list.* Spend a few minutes discussing each environment and answering questions so that, by the end of the hour, students have a basis for deciding which habitat they would prefer to study. During the last ten minutes of the hour students gather in their small groups. Each group must hand in a list of its first, second, and third choices of environments to study.

Note:

- "Natural Environments" was designed to allow students to study various kinds of animal and plant habitats in their home state. Therefore, before the project begins, you must prepare a list of ten or more natural environments that can be found in your state. This project was developed in Michigan, so here is a sample list of ten Michigan environments that were actually used with students in a classroom:

  Open plain or prairie
  Deciduous forest
  Coniferous forest
  Pond
  Slow-moving river
  Fast-moving stream
  Cedar swamp
  Great lakes/dunes
  Marsh or bog
  Small lake

HOUR 3: Tell students which of their environment choices their group will be allowed to study; the method for determining this is up to you. Then students meet in small groups, and each group develops a "Natural Environment Roster." (Students were given this handout during Hour 1.) Information sharing should be encouraged. It is crucial that *plenty* of resource material be available at all times for this project. If possible, arrange to conduct the research portion of this project in the school library or even in a public library if that option is open to you. During this hour students try to identify the most common plants and animals that live in the environments they have chosen to study.

Notes:

- After groups know which environment choice they will be allowed to study (each group should have a *different* environment), their first assignment is to create a "Natural Environment Roster." This is more difficult than it sounds because there is no single resource that lists all of the plants and animals in a particular environment in any given state. Often students must make educated guesses about which animals live in a specific environment and look these animals up in various sources to verify that they do live in that environment. From these sources they will find references to other plants and animals. This process can be time-consuming and at times, frustrating. To reduce frustration it is *crucial* that plenty of reference material be available.

- Encourage students to help one another as they create their environment rosters. If a student in one group finds that a certain animal lives in an environment that another group is studying, the information should be shared. If you have 25 investigators studying animal books and sharing information, the environment rosters will take shape quickly, and be fairly complete.

- There is, obviously, a great deal of overlapping in the animals found in various environments. This is another chance to get students to share and cooperate by supplying information to one another.

HOUR 4: Students continue to work on their environment rosters.

HOUR 5: Students complete their environment rosters by the end of this hour. Those groups who finish early and claim they are "done" should be encouraged to spend more time discovering additional roster entries. It is literally impossible to list all the life forms in any habitat in three hours, given your students' knowledge of the subject. More time can always be well spent looking through another book or finding new leads. Rosters are handed in at the end of the hour.

HOUR 6: Groups get their environment rosters back, and each student decides on two plants, two animals, or one of each to study. Group members meet to choose topics and discuss what is needed to produce a quality mural. Group decision-making plays a large role in each student's selection of plants and/or animals. At the end of this hour each group must hand in an "Animal and Plant Selections" sheet. Students were given this handout during Hour 1.

HOUR 7: Return the selection sheets; students can now begin to study the plants and animals they have chosen. They should gear most of their research toward finding information about the areas listed on the "Natural Environments

Research Guide." Students were given this handout during Hour 1. All research information is to be recorded on notecards.

Note:

- Students can help one another with research. One student may say, "Is anyone studying white-tailed deer? I found a good book about them." This type of sharing should be encouraged as long as each person is doing his or her own research.

HOURS 8-13: Students continue to study their plants and animals and write research papers.

HOUR 14: Research papers are turned in at the beginning of the hour. Students then assemble in their groups to begin planning murals. At the end of the hour each group hands in a rough sketch of its mural, showing where animals and plants will be located and indicating other prominent features to be included.

Notes:

- When research papers are handed in this hour you must make a decision about grading them:

  a. If you grade them carefully, for such things as accuracy, effort, neatness, and sufficient information, this takes time and Hour 15 should not be scheduled until you are ready to hand back all of the reports.
  b. If you want Hour 15 to come immediately after Hour 14, you can check each report quickly, mark whether it is "satisfactory" or "unsatisfactory," and have all reports ready to hand back the next day so students can begin work on murals. In this case, reports should be handed back in at the end of the course to be graded more carefully.

- Be sure to require a *plan,* or a sketch, of the mural each group intends to produce; it should be handed in to you and receive your approval before work on the mural begins.

HOURS 15-19: Return the checked research reports and rough sketches at the beginning of Hour 15. Students work in small groups on their murals until they are completed. There is no reason for students to be idle: a mural can accept infinite detail, so there is always a flower to be added, a butterfly to color, an empty space that could be filled with another tree, and so forth. Often, students who act like they don't want to work may be having trouble getting along in a group. Be aware of this and help groups work through these rough spots. Everyone's contribution is important.

When the murals are finished (Hour 19), display them in the room or elsewhere in the school. Tell students to complete the "Self-Evaluation Sheet" that was handed out during Hour 1. Students evaluate themselves and their fellow group members on how well they worked together during this project. It is up to you to decide how you want to determine a final grade for "Natural

Environments." A "Final Evaluation" form is provided with this project and there are also several evaluation sheets in the Teacher's Introduction to the Research Guide which you may find useful for grading student work.

HOUR 20: Conduct a class discussion focused on how higher level thinking skills were applied during this project, especially at the synthesis level. Information was drawn from diverse places, processed by individuals into reports, and then organized on a mural according to a *plan*. It is pointed out that most organizations and businesses deal with information in the same way the small groups did. Individuals are given responsibilities and then each person's effort is blended into a final product. These are key verbs to use when discussing synthesis:

Design—students designed their own parts of the mural.

Plan—the murals were planned by group discussion and consensus.

Create—the murals were created from originally unconnected facts and the combined imaginations of the students.

General Notes About This Project:

- This project can be shortened by assigning some of the work done in Hours 7 through 13 as homework. These are the classroom hours spent on research and writing reports.

- The Teacher's Introduction to the Student Research Guide and the Student Research Guide itself offer several evaluation forms, informational handouts, and checklists that may be useful. Take the time to look over these materials. Of particular value for this project are evaluation forms for notecards and posters (or murals), handouts on notecards and bibliographies, and a daily log in which students keep track of their work.

Name _____ Date _____

NATURAL ENVIRONMENTS
Student Assignment Sheet

Natural environments are places that have not been greatly affected by human beings, where animals and plants live together in an ecosystem. The study of natural environments provides an excellent opportunity for independent learning, since there are so many different habitats and such a wide variety of wildlife to investigate. Learning on your own about a specific natural environment in your state will help you understand more clearly how animals and plants live: their physical needs and their reliance upon the never-ending life cycle that is called the balance of nature.

Here is the Natural Environments assignment.

I. Your teacher has provided a list of natural environments that can be found in your home state. Write this list in the spaces below:

Ten Natural Environments Found in My State

1. _____   6. _____

2. _____   7. _____

3. _____   8. _____

4. _____   9. _____

5. _____   10. _____

II. After discussing all ten environments in class, meet with your group and choose three environments that you would like to study. List them in order of preference below

A. _____

B. _____

C. _____

Your teacher will decide which natural environment each group will be allowed to study. Every attempt will be made to assign your group one of its three choices but there are no guarantees, since no two groups will study the same environment.

III. When you find out which environment your group will study, record it on the line below:

_____

IV. *As a group,* find as much information as possible about this specific environment.

    A. Make a list of animals that live in it.

    B. Make a list of plants that live in it.

    C. Put all of these animals and plants on your group's "Natural Environment Roster."

    D. Turn the completed roster in to be checked.

V. Each person in the group must select at least two plants, two animals, or one plant and one animal from the environment roster to study as an individual project. Record these choices on your group's "Animal and Plant Selections" sheet.

    A. Turn the completed selection sheet in for approval.

    B. Be sure that the plants and animals your group has chosen fit into a food chain or web.

VI. Individual Research

    A. Follow the "Natural Environments Research Guide" as you prepare to write reports about your two plants or animals.

    B. Write one report for each plant or animal.

       1. Each written report should be at least one full page long.

       2. Include at least one drawing of the plant or animal, neatly labeled and showing it in its natural environment.

       3. Record all of the information from your research on notecards and refer these to a bibliography. Notecards and a bibliography will be turned in with reports.

    C. Reports will be handed back before murals are started.

VII. Using information from the reports, your group will design and create a wall mural showing how these animals and plants live together in the environment.

    A. Make a small drawing of what the mural will look like, and turn it in to be checked. This is your plan, and it will serve as a blueprint, so put some detail into it.

    B. When your group is satisfied with its plan and the teacher has okayed it, begin working on the mural. Take your time and be neat. Put as much quality into it as possible. *Sketch drawings in pencil before coloring them.*

    C. You will evaluate each other when the mural is finished. The evaluation will cover ten areas:

1. Total contribution to the project	6. Effort
2. Sharing ideas	7. Willingness to work
3. Accepting other's ideas	8. Quality of work
4. Concentrating on the project	9. General attitude
5. Taking care of materials	10. Organization and neatness

NATURAL ENVIRONMENT ROSTER

Natural Environment: _____

Group Members: 1._____   4._____

2._____   5._____

3._____

The first step in studying a natural environment is to find out what plants and animals live in it by looking through books, encyclopedias, magazines, and other resources. Use this handout to list as many animals and plants as you can that fit into these categories and are found in the environment. Put *O* for omnivore, *C* for carnivore, or *H* for herbivore after each animal. Put a *D* beside decomposers and a *G* beside green plants. If you do not understand any of these terms, ask questions in class or look them up before completing this handout.

1. Mammals

2. Birds

3. Reptiles

4. Amphibians

5. Fish

6. Insects

7. Invertebrates (other than insects)

8. Trees

9. Shrubs and bushes

10. Grasses and other perennials

11. Annuals

12. Fungi, algae, ferns

13. Other life forms

Name _____ Date _____

ANIMAL AND PLANT SELECTIONS

Environment: _____

The next step in this project is for everyone to choose two species to study. From the roster, each student must choose a plant and an animal *or* two plants *or* two animals to study. On this handout, record "who" chose "what" to study individually. These choices should fit into a food web, so make sure the list includes a variety of predators, prey, green plants, and decomposers. Turn these selections in for approval before beginning your research. It may be necessary for someone to study an *order* or a *family* instead of an individual species if information is not available about specific plants or animals.

Name of Plant or Animal (or family)	Person Studying
1.	
2.	
3.	
4.	
5.	
6.	
7.	
8.	
9.	
10.	
11.	
12.	
13.	
14.	
15.	
16.	
17.	
18.	
19.	
20.	

NATURAL ENVIRONMENTS RESEARCH GUIDE

This handout is designed to help you find information about the species you have chosen to study. The items on this list provide an excellent outline for the written report as well. As you find information, record it on notecards first, and then write a one- to two-page report for each animal or plant that you study. This information will be used when your group plans its mural.

1. Scientific name: Common name:

2. Physical description: Include a drawing.

3. Close relatives: Name other animals/plants within its family or genus.

4. Diet: Is the animal/plant a decomposer, green plant, carnivore, omnivore, or herbivore? What are its food or nutrient sources?

5. Adaptations: What does your plant or animal have that helps it survive in its environment? Show them on the physical description drawing.

6. Seasonal changes: How does your plant or animal adjust to different seasons, and why?

7. Food web: Describe a food web that includes your plant or animal. Work with other group members to develop this for your mural.

8. Reproduction: Investigate such areas as mating behavior, care of young, social structure, special territory needs, pollination, seeds, and seed distribution.

9. How is it affected by humans? Is it endangered? Has its territory (range) been reduced? Is pollution affecting it? Is it hunted? Is it used as food?

10. Bibliography

Name _____ Date _____

NATURAL ENVIRONMENTS
Self-Evaluation Sheet

List the people in your group (including yourself) on the spaces across the top of this page. As fairly and honestly as you can, give each person an evaluation for the ten categories itemized in the left-hand column. Circle a number for each category under each group member's name: "1" is poor, "5" is excellent.

Names of Group Members

Evaluation Categories					
1. Total contribution to the project	1 2 3 4 5	1 2 3 4 5	1 2 3 4 5	1 2 3 4 5	1 2 3 4 5
2. Sharing ideas	1 2 3 4 5	1 2 3 4 5	1 2 3 4 5	1 2 3 4 5	1 2 3 4 5
3. Accepting others' ideas	1 2 3 4 5	1 2 3 4 5	1 2 3 4 5	1 2 3 4 5	1 2 3 4 5
4. Concentrating on the project	1 2 3 4 5	1 2 3 4 5	1 2 3 4 5	1 2 3 4 5	1 2 3 4 5
5. Taking care of materials	1 2 3 4 5	1 2 3 4 5	1 2 3 4 5	1 2 3 4 5	1 2 3 4 5
6. Effort	1 2 3 4 5	1 2 3 4 5	1 2 3 4 5	1 2 3 4 5	1 2 3 4 5
7. Willingness to work	1 2 3 4 5	1 2 3 4 5	1 2 3 4 5	1 2 3 4 5	1 2 3 4 5
8. Quality of work	1 2 3 4 5	1 2 3 4 5	1 2 3 4 5	1 2 3 4 5	1 2 3 4 5
9. General attitude	1 2 3 4 5	1 2 3 4 5	1 2 3 4 5	1 2 3 4 5	1 2 3 4 5
10. Organization and neatness	1 2 3 4 5	1 2 3 4 5	1 2 3 4 5	1 2 3 4 5	1 2 3 4 5
TOTAL (50 pts. possible)	_____	_____	_____	_____	_____

Now record each group member's name (including your own) on a space below, and write what you think that person's greatest contribution to the "Natural Environments" project was.

Group Members   **Greatest Contribution**

1. _____ : _____
   (Your Name)

2. _____ : _____

3. _____ : _____

4. _____ : _____

5. _____ : _____

Record any other comments you have about this project on the back of this page.

Name ———————————————— Date ————————————————

NATURAL ENVIRONMENTS
FINAL EVALUATION

I. Topic 1 ————————————————————————

   A. Bibliography   ————— (0–5 pts.)

   B. Notecards   ————— (0–5 pts.)

   C. Drawing   ————— (0–5 pts.)

   D. Report

      1. Grammar, spelling, paragraphs   ————— (0–5 pts.)
      2. Content (appropriate information) ————— (0–5 pts.)
      3. Neat, easy to understand, and written in your own words ————— (0–5 pts.)

II. Topic 2 ————————————————————————

   A. Bibliography   ————— (0–5 pts.)

   B. Notecards   ————— (0–5 pts.)

   C. Drawing   ————— (0–5 pts.)

   D. Report

      1. Grammar, spelling, paragraphs   ————— (0–5 pts.)
      2. Content (appropriate information) ————— (0–5 pts.)
      3. Neat, easy to understand, and written in your own words ————— (0–5 pts.)

III. Mural (environment: ————————————————)

   A. Worked with others   ————— (0–10 pts.)

   B. Contributed to the finished mural ————— (0–10 pts.)

   C. Took care of materials   ————— (0–5 pts.)

   D. Showed effort   ————— (0–5 pts.)

   E. Was neat   ————— (0–5 pts.)

   F. Shared ideas and was willing to compromise ————— (0–5 pts.)

                 TOTAL ————— (100 points possible)

INDEPENDENT STUDY PROJECTS

Teacher Preview

Project Topics:

Plant Science

Planning a Garden

Astronomy

Geology: Minerals

Sleeping and Dreaming

Mammals of North America

Whales

Animal Signs

Body Systems

Forestry: Tree Identification

General Explanation: There are several ways of presenting these projects, from giving students the handouts and a due date three weeks later to making daily lesson plans for presenting information and allowing students to work on their projects in class. The hour-by-hour scheduling of these projects is left to you with one word of caution: be sure to provide plenty of time for completion of the projects, regardless of the method used in presenting them. Also, try to provide a forum for students to exhibit or display their work, such as parent conferences, classroom presentations, open houses, school-wide presentations, small displays, or evening programs.

Length of Each Project:

1. 2 classroom hours

2. 3 weeks of students' own time

Level of Independence: Advanced

Goals:

1. To provide a choice of subjects for students to study.

2. To require the application of independent learning skills.

During This Project Students Will:

1. Choose projects to work on.

2. Follow brief outlines to complete project requirements on their own.

3. Apply independent learning skills.

4. Design ways to present what is learned.

Skills:

Preparing bibliographies

Collecting data

Accepting responsibility

Concentration

145

Library skills

Listening

Making notecards

Observing

Summarizing

Grammar

Handwriting

Neatness

Paragraphs

Sentences

Spelling

Organizing

Outlining

Setting objectives

Selecting topics

Divergent-convergent-evaluative thinking

Following and changing plans

Identifying problems

Meeting deadlines

Working with limited resources

Controlling behavior

Following project outlines

Individualized study habits

Persistence

Sharing space

Taking care of materials

Time management

Personal motivation

Self-awareness

Sense of "quality"

Setting personal goals

Creative expression

Creating presentation strategies

Diorama and model building

Drawing/sketching/graphing

Poster making

Public speaking

Self-confidence

Teaching others

Writing

Handouts Provided:

- "Introduction to the Independent Project"
- "Student Assignment Sheet" for each area of study
- Teacher's Introduction to the Student Research Guide (optional; see Appendix)
- Student Research Guide (optional; see Appendix)

PROJECT CALENDAR:

HOUR 1: _____ Students choose their projects and are given assignment sheets. HANDOUT PROVIDED	**HOUR 2:** _____ Students present or turn in their projects. STUDENTS TURN IN WORK	**HOUR 3:** _____
HOUR 4: _____	**HOUR 5:** _____	**HOUR 6:** _____
HOUR 7: _____	**HOUR 8:** _____	**HOUR 9:** _____

Lesson Plans and Notes

HOUR 1: The projects in this section of the book are presented as independent learning projects, to be done outside the classroom on the students' own time. They are not necessarily suitable for every student in class; there may be only a few who are allowed to produce projects on their own. During Hour 1 introduce students to the topics available and hand out assignment sheets. Establish due dates also. Students are expected to design their projects and produce them within that time frame.

Note:

- Project outlines may be altered or rewritten to meet specific needs. They can be used as totally independent projects that are done at home, individualized projects that are done in the classroom, small-group projects, or full-group projects that are worked on as a class.

HOUR 2: Students present or turn in their projects.

General Notes About These Projects:

- Each of the projects is self-explanatory, but *the handouts are brief and make the assumption that the students who are using them have previous experience in working from project outlines.* The projects are designed to allow students to decide such things as methods of recording information, the number of information sources, methods of presentation, and types and lengths of reports.

- These projects are written in the form of "learning contracts." It is not necessary to require student signatures, but this is an especially good idea if work is to be done independently outside the classroom.

- The Teacher's Introduction to the Student Research Guide (see Appendix) offers several evaluation forms that may be used at the conclusion of a project. The Student Research Guide (Appendix) provides instructional handouts, several useful checklists, and a daily log that may be incorporated into project requirements. It is suggested that, above all, the daily log be used, since it is a convenient way for students to keep track of their own progress and record their daily activities.

Name ——————————————— Date ———————————

INTRODUCTION TO THE INDEPENDENT PROJECT

The ultimate goal of a scientist, or a student of science, is to be allowed to choose a subject for independent study. Some of the greatest discoveries of all time were the result of individuals working by themselves to learn about the nature of things. These people used the knowledge of others to help them create something new. Such people as Thomas Edison, George Washington Carver, Albert Einstein, Sir Isaac Newton, Eli Whitney, Copernicus, Henry Ford, Jonas Salk, Madame Curie, and many others were all independent learners. They made their great contributions because they knew how to design a project and work on it until it was completed. You may or may not be another Einstein, but regardless of your future as a scientist, you can benefit from an independent study project on a science topic.

The latter part of the twentieth century offers an astonishing number of topics to explore in the field of science, and each year brings more discoveries and more things about which to learn. A school cannot possibly offer courses in all the areas of science that are now opening up, and a general course of study doesn't even have time to *mention* them all. How is a student to learn about areas of science that are not offered in school? The answer, obviously, is to study topics of interest independently, thereby determining for him- or herself what to learn.

The topic choices provided with this project are not taken from the more difficult physical sciences like physics and chemistry, which require background information. They are topics that have plenty of information available and that are interesting to study. As you work on the project keep in mind that it is preparing you for future, more complex, explorations in science. The skills you learn to use now will be skills that are at your disposal in future years when you feel the need to learn about something that isn't offered in school or isn't covered in enough depth to suit your needs.

INDEPENDENT PROJECT TITLES

Plant Science

Planning a Garden

Astronomy

Geology: Minerals

Sleeping and Dreaming

Mammals of North America

Whales

Animal Signs

Body Systems

Forestry: Tree Identification

Name _____ Date _____

PLANT SCIENCE
Student Assignment Sheet

There are many interesting plants that can be studied to learn about the plant kingdom in general. This project allows you to learn about one plant species.

 I. Choose a plant to study in depth.

 II. Focus your research on the following areas:

  A. *History*—Origin, native area where it was raised, gathered, or used, historical significance, and so forth.

  B. *Scientific information*—Name (common and scientific), family the plant belongs to and information about the family, identification characteristics, basic plant structure, botanical information, and so forth.

  C. *Growth*—How the plant is started (seed, rooting, cutting), conditions necessary for good growth (light, water, nutrients, soil), size, rate of growth, life span, and a description of the plant.

  D. *Uses*—Food, medicine, superstitions, dyes, scents, decorations, and so forth. Also, if it grows naturally in the wild, how does it fit into nature's balance?

 III. Use these questions as a research guide. Provide a bibliography.

  A. How long has man known about this plant?
  B. Where was it first discovered, and by whom?
  C. Was it imported from a foreign country? When? Where? Why?
  D. Is the plant a pure strain or is it a hybrid? What is a hybrid?
  E. How many hybrids of this plant exist?
  F. Are there any special uses for this plant? What are they?
  G. Is your plant a vegetable, a fruit, or neither?
  H. What type of soil and climate is the best for this plant?
  I. What color varieties does this plant have?
  J. How does it propagate itself?
  K. What is its common name?
  L. What is its scientific name?
  M. Provide a physical description of the plant. Include detailed drawings.
  N. What are its close relatives?
  O. What seasonal changes does it go through?

Optional

IV. Start, raise, and observe a plant.

　A. Start the plant from a seed, stem cutting, leaf cutting, or whatever way is appropriate. Certain plants must be started in a particular way. Make sure you find out which way is best.

　B. Observe the plant for at least six weeks. Record daily growth, conditions present (temperature, humidity, available light, and so forth), how often you water it, how much water you give it, when you turn it, and any other observations you can think of. Be as precise as possible.

　C. Make a chart of your day-to-day observations. Include any conclusions you can draw from the data.

　D. It is a good idea to start more than one plant.

　　1. This will help ensure a live, healthy plant for your project.

　　2. With more than one plant you can expand the project by varying the amounts of light, water, soil, nutrients, and warmth; carefully record all of your observations.

　E. Make pressings of leaves, flowers, or the entire plant, to include with the research report. Photographs are also a nice addition to a research report.

_____ has completed all assignments, is passing all subjects, continues to do satisfactory work, and is allowed to begin an independent project titled "Plant Science."

Student _____

Teacher _____

Date signed _____

Due date _____

PLANNING A GARDEN
Student Assignment Sheet

Gardening is a popular hobby in America and represents one of the most important undertakings in the world: raising food. This project allows you to find out how gardens are planned and cared for. It might even get you started on a hobby that will last the rest of your life.

I. You are to design a hypothetical but practical garden.

A. Size—The size is up to you. Make sure it is big enough to be valuable as a food supply.

B. Plants to include—Make sure you have a variety. Include at *least* five of these types of plants in your garden:

1. Legumes (beans and peas).

2. Greens (such as lettuce, mustard greens). Remember a little goes a long way.

3. Vining plants (cucumbers, squash, pumpkins, gourds, melons).

4. Root plants (beets, carrots, radishes, onions, potatoes).

5. At least two herbs.

6. Flowers—they are an eye-pleasing addition to a garden and certain flowers are beneficial as insect deterrents.

7. Cabbage family (cabbage, cauliflower, broccoli, Brussels sprouts).

8. Other common garden plants (tomatoes, corn, peppers, strawberries).

C. Arrangement—Decide the direction of rows (running east and west or north and south) and the location of various kinds of plants in the garden. Arrange plants so they are not overcrowded and so that compatible plants are next to one another. Also, don't overlook esthetic value—flowers and an orderly arrangement enhance a garden's looks.

D. Timing: Create a planting schedule (make a chart). Plants should be started as early as possible in a productive garden. "Possible" means the best time for a particular plant. To ensure a continuous supply of food all summer long, two or three plantings should be planned at three-week intervals. You will need to consider the maturity time for each plant and the growing season in your part of the country.

II. Drawing—Make a scale drawing of your garden plot. Show the arrangement of plants, number of plants, time when they are planted, and so forth.

III. Report—Explain the value of a garden and support your decision for size, plant types, arrangement, and timing. Predict how much *yield* you expect from the garden. Your report should also consider any problems that may be encountered in raising plants, such as destructive insects, disease, rodents, poor soil, lack of water, insufficient sunlight. Find out how you would handle each problem. Be as detailed as you can; also be practical.

IV. If you are interested in a more scientific approach to this project, focus your research on genetics: heredity and hybrids. A good place to start is with the work of Gregor J. Mendel. Other possible areas of research are soil composition, fertilizer, herbicides, and insecticides.

V. Resources—You can visit greenhouses and plant shops. There are also many good books and magazines on gardening. Don't overlook the gardeners that you or your parents know. They are experts and possibly even outstanding in their field, so to speak.

VI. If possible, plant a garden at home.

_____ has completed all assignments, is passing all subjects, continues to do satisfactory work, and is allowed to begin an independent project titled "Planning a Garden."

Student _____

Teacher _____

Date signed _____

Due date _____

© 1987 by The Center for Applied Research in Education, Inc.

Name _____ Date _____

ASTRONOMY
Student Assignment Sheet

The purpose of this project is to study one topic within the area of astronomy. Choose a topic from the list below or create your own. Here are the project requirements.

I. Design a visual display that illustrates important concepts or facts about the topic.

II. Write a report, make a cassette tape, or give an oral presentation that teaches or presents what you have learned from your research.

List of Possible Topics

1. Spectrum—color, electromagnetic, and radio: their uses in the study of astronomy
2. Stars—brightness, color, size, distance, constellations, composition
3. Measurements in space—light years, relative distances, space warps
4. Speed of light—what it means in time and distance
5. Binary stars
6. Pulsars
7. Quasars
8. Black holes
9. Nebulas
10. Radio astronomy
11. Telescopes and observatories
12. Famous astronomers and their discoveries
13. Kepler's laws of planetary motion
14. Mapping space: A system of coordinates
15. Van Allen belt
16. The planets
17. The sun
18. The moon
19. Satellite explorations of the planets
20. Other _____

© 1987 by The Center for Applied Research in Education, Inc.

_____ has completed all assignments, is passing all subjects, continues to do satisfactory work, and is allowed to begin an independent project titled "Astronomy."

Student _____

Teacher _____

Date signed _____

Due date _____

Name _____ Date _____

GEOLOGY: MINERALS
Student Assignment Sheet

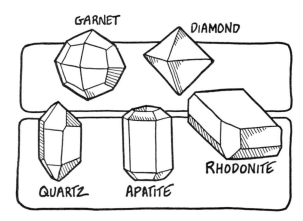

Geology is the study of the "stuff" our planet is made of: rocks and minerals. The purpose of this project is to study minerals that are of interest to you.

I. Mineral Collection

    A. Collect and identify at least six different minerals.

    B. Write a brief report, including how each mineral was formed, its uses, its physical characteristics, its chemical makeup, where it is found naturally, and so forth.

    C. Make a display of these minerals and set it up in the classroom. Interesting facts from your report should be presented on the display.

II. Precious Minerals (gemstones)

    A. Choose a precious mineral and write a report about it:
       1. Where is the mineral found?
       2. Why is it precious?
       3. How is the mineral made into a gem? (cutting, polishing, faceting)
    B. Make a list of minerals that are considered to be precious.

III. Diagram the crystalline structure of two or three minerals from your collection.

IV. Explain what "cleavage" is in a mineral.

_____ has completed all assignments, is passing all subjects, continues to do satisfactory work, and is allowed to begin an independent project titled "Geology: Minerals."

                      Student _____

                      Teacher _____

                      Date signed _____

                      Due date _____

Name _____ Date _____

SLEEPING AND DREAMING
Student Assignment Sheet

Sleep and dreams have been studied since ancient times. You will not be interpreting dreams, but you will learn about the nature of sleep. The following list provides suggestions for your research. If there are other areas you want to study, feel free to do so.

1. Research the four stages of sleep, and what your body is doing in each one. When do nightmares occur?
2. What is REM and what occurs during REM sleep? What studies have been done on REM?
3. What happens to REM-deprived animals? humans? What are some common ways people deprive themselves of REM?
4. What happens if a person is totally deprived of sleep?
5. Find out what an EEG machine is and what it measures.
6. How much sleep do people need? What purposes does sleep serve (physical and mental)?
7. Research theories about dreams and what they mean.
8. Here are some other areas you may wish to investigate: snoring, insomnia, movement during sleep, sleepwalking, breathing patterns, naps, narcolepsy, sleep studies, and experiments.

Write a report about sleeping and dreaming, and include a "dream" log. This will involve writing down your dreams every morning when you wake up. Do this over a two-week period. Keep a pencil and paper right next to the bed, so you can write down your dreams *before* you get out of bed. Otherwise you will forget them! These morning notes are merely a rough draft. Rewrite them into a well-written paper and combine it with the report on sleep and dreams.

_____ has completed all assignments, is passing all subjects, continues to do satisfactory work, and is allowed to begin an independent project titled "Sleeping and Dreaming."

Student _____

Teacher _____

Date signed _____

Due date _____

© 1987 by The Center for Applied Research in Education, Inc.

Name _____ Date _____

MAMMALS OF NORTH AMERICA
Student Assignment Sheet

© 1987 by The Center for Applied Research in Education, Inc.

The purpose of this project is to study one mammal of North America in detail.

I. Build a shoebox diorama, make a mural, build a mobile, or make a poster that shows the mammal in its natural environment. An example is a mural showing a beaver's habitat: stream, pond, dam, den, and surrounding forest.

II. Write a report that tells about the animal's habits and behavior, adaptations, food, habitat, close relatives, natural enemies, classification, geographic territory (range), social structure, communication, reproduction, and the like.

III. Make a poster that shows your North American mammal in a food chain and a simple food web.

IV. You may also want to find out if the animal is endangered, how its territory is affected by man, and what its population distribution is.

_____ has completed all assignments, is passing all subjects, continues to do satisfactory work, and is allowed to begin an independent project titled "Mammals of North America."

Student _____

Teacher _____

Date signed _____

Due date _____

Name _____ Date _____

WHALES
Student Assignment Sheet

For this project you will study the "giant mammals." Collect and record information as you answer as many questions about them as you can. Share your information with the class by presenting your findings in a detailed report that includes some type of visual presentation.

   I. Answer these general research questions. (These are only examples; think of others!)

      A. How many species of whales exist at the present time?

      B. How many have become extinct?

      C. What is the largest whale (length and weight)?

      D. What is the smallest whale (length and weight)?

      E. How are whales classified (phylum, class, order)?

      F. How do whales communicate with each other?

      G. How many miles do whales swim when they migrate?

      H. Do all whales migrate?

      I. What do whales eat? Do all whales eat the same things?

      J. How deep can a whale swim underwater and how long can it stay under?

      K. Why were whales hunted 150 years ago? Why are they hunted today?

      L. What size are whales when they are born? How long is the gestation period? How many babies are born at one time?

      M. What is "baleen"?

  II. Choose a specific species and find as much information as you can about it.

_____ has completed all assignments, is passing all subjects, continues to do satisfactory work, and is allowed to begin an independent project titled "Whales."

           Student _____

           Teacher _____

           Date signed _____

           Due date _____

© 1987 by The Center for Applied Research in Education, Inc.

Name _____ Date _____

ANIMAL SIGNS
Student Assignment Sheet

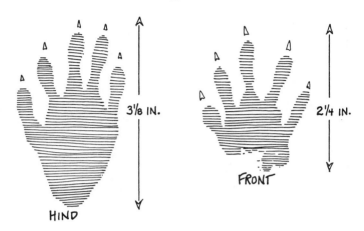

Animal signs are signals left behind by animals. By studying the signs it is possible to identify what kind of animal made them. Hunters and trackers have used this skill for centuries, but it is also a fascinating way for anyone to learn about the habits and characteristics of wildlife.

I. Find and read a good book about animal signs. Animal signs include tracks, markings, scats, homes, and sheddings, among other things.

II. After reading the book, choose twenty animal signs and make sketches of them on a posterboard. Your poster will look like this:

sketch	sketch	sketch	sketch	sketch
sketch	sketch	sketch	sketch	sketch
sketch	sketch	sketch	sketch	sketch
sketch	sketch	sketch	sketch	sketch

III. Draw your sketches in such a way that the entire poster can be matted to give it a professional look. The matte is a piece of colored posterboard the exact size of your white posterboard, with pieces cut out so that your sketches will be neatly framed:

Directly below each sketch, on the matte, identify the animal and its sign. Your matte will look like this:

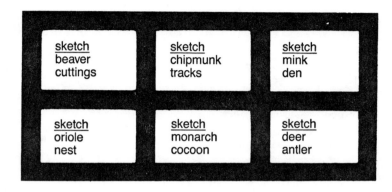

IV. Choose one of your twenty animal signs and make a large drawing of it on posterboard. Label everything and include as much information as possible.

V. Make your sketches as neat, accurate, and informative as possible. Printing should also be neat and easy to read. Your poster will be displayed so put as much quality into it as you can.

_____ has completed all assignments, is passing all subjects, continues to do satisfactory work, and is allowed to begin an independent project titled "Animal Signs."

Student _____

Teacher _____

Date signed _____

Due date _____

Name _____ Date _____

BODY SYSTEMS
Student Assignment Sheet

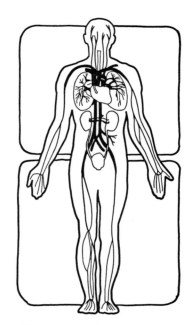

In this project you will learn about the different systems of the body. One to six people can work on this project (one person per system). To learn about the body systems and how they work you will do the following.

I. Make full-scale drawings of the body. Put a strip of white rollpaper, as long as you are, on the wall. By shining a light from a projector onto the paper, and standing in front of it, someone else can draw your body outline.

II. Study at least one system for each person in your group. (If you are working alone, choose any system you want.) Do one system per drawing; if you decide to do the respiratory system and the nervous system, then make two body outlines and turn in two system drawings.

III. Write a report about the body system you have chosen to study.

IV. Within the body outline, you can draw the appropriate parts of the body, depending upon which particular system you are studying. You can choose from the circulatory system, respiratory system, digestive system, central nervous system, skeletal system, and the lymphatic system. Carefully label each illustration with brief explanations about how the system works.

V. You will present your work to the class.

_____ has completed all assignments, is passing all subjects, continues to do satisfactory work, and is allowed to begin an independent project titled "Body Systems."

Student _____

Teacher _____

Date signed _____

Due date _____

Name _____ Date _____

FORESTRY: TREE IDENTIFICATION
Student Assignment Sheet

For this project you will study one aspect of forestry: tree identification. The project has two parts: one is a presentation and the other is a collection. You *must* do both parts to complete the assignment.

I. Presentation

    A. Choose a tree to study. It should be a species that can be found around your home, school, or community.

    B. Write a report about the tree. It should include:

        1. Scientific and common name

        2. Description of leaf, branch, bud, flower, bark, seeds, wood, shape, and size

        3. Habitat

        4. Distribution

        5. How the tree is used by man

        6. A bibliography

    C. Make a pencil sketch of the leaves and twigs. Have good examples to look at while you draw.

    D. Make a display that can be exhibited in the classroom. This display should show examples or drawings of leaves, twigs, buds, bark, flowers, seeds, habitat, and the like.

II. Collection (21 trees)

    A. Collect two *very* good examples of leaves from the tree you are studying and press them. Mount these samples on notebook-size paper and include them with your report.

    B. Choose 20 different kinds of trees, collect leaves from each, press the leaves, and mount them on notebook-size paper. Label them with their common names. Scientific names may also be included, along with any other information you wish to present.

Optional

    C. If you want to focus more closely on the science of forestry, you may wish to include some of these topics in your report:

1. Photosynthesis	4. Diseases	6. Uses of forest products
2. Transpiration	5. Stages of growth	7. Forest management
3. Osmosis		

_____ has completed all assignments, is passing all subjects, continues to do satisfactory work, and is allowed to begin an independent project titled "Forestry: Tree Identification."

Student _____

Teacher _____

Date signed _____

Due date _____

SCIENCE FAIR OR OPEN HOUSE

Teacher Preview

Length of Project: 4 hours, plus one afternoon or evening.

Level of Independence: Advanced

General Explanation:

The Science Fair (or Open House) is designed to let students present their work to parents, relatives, friends, and students from other classes. It is a way of culminating a science curriculum so that all of the units, lessons, and projects coalesce into one final presentation by an entire class.

Goals:

1. To allow students an opportunity to display their knowledge and understanding of science.
2. To place emphasis on independent learning.
3. To provide a project that unifies the year's curriculum.

During This Project Students Will:

1. Choose areas of the curriculum they would like to present.
2. Develop presentations and displays.
3. Present their work to audiences of parents, teachers, and other students.

Skills:

All of the skills on the Skills Chart at the front of the book may be used as a guide.

Handouts Provided:

Any of the handouts in this book may be used.

PROJECT CALENDAR:

HOUR 1: _____	HOUR 2: _____	HOUR 3: _____
Introduction to the Science Fair or Open House. Discussion about different areas of the curriculum and how they could be presented to parents.	Students bring in presentation ideas and these are discussed before assignments are made at the end of the hour. PREPARATION REQUIRED	Presentation descriptions are discussed and turned in. STUDENTS TURN IN WORK
HOUR 4: _____	HOUR 5: _____	HOUR 6: _____
Dress rehearsal for the Science Fair (or Open House). NEED SPECIAL MATERIALS	For an afternoon or an evening students present examples of the kinds of things they learned all year in their science class.	
HOUR 7: _____	HOUR 8: _____	HOUR 9: _____

Lesson Plans and Notes

HOUR 1: Introduce students to the idea of a science fair or open house that is designed to present the year's curriculum to parents and other interested people. Spend the hour discussing the various subjects that were covered during the year and asking for ideas about the best ways to present those subjects to parents. Students are told to come to the next class prepared with at least three ideas for science fair/open house presentations, based upon the year's science curriculum.

HOUR 2: Spend the hour listing the ideas students have for science fair/open house presentations. Write each student's ideas on the board for easy reference when, at the end of the hour, assignments are made. Make every attempt to give students one of the three ideas they brought to class. For the next hour, each student is to write a description of his or her presentation.

Notes:

- It is a good idea to bring some presentation ideas to class this hour, especially for areas of the curriculum that students may not choose. The open house should cover all parts of the year if possible. Also, some students may quibble over a particularly choice idea that more than one person has thought of, and arbitration is more successful if there are alternatives to offer.

- Here are some presentation ideas, based upon this book:

    a. A presentation of the terms "load," "work," and "force," followed by a demonstration of an inclined plane
    b. Same for levers
    c. Same for pulleys
    d. An experimental setup showing how "Speed of Toboggans" is conducted
    e. A discussion of atomic structure with drawings, definitions of terms, and explanations
    f. A presentation of the periodic table of the elements: what it shows and how it can be used
    g. An explanation of ionic bonding: drawings and lecture
    h. An explanation of how to complete acid-base equations
    i. An experiment: copper plating, along with an explanation of why it works
    j. A lecture on animal taxonomy, with emphasis on the five classes of vertebrate animals
    k. An oral presentation on a specific animal, along with a visual display
    l. An oral presentation on a specific insect, along with a visual display
    m. An oral presentation on a specific ocean animal, along with a visual display

n. A group presentation of a natural environment, with discussion of animals, plants, habitat, and food web

o. A description of how a shoebox diorama can be built to show animals in the wild, and a display of such shoeboxes

p. A display of imaginary animals and an explanation of animal adaptations

q. Presentations on any or all of the independent study projects

r. An explanation of independent learning skills and why they are important in a science class

HOUR 3: Students describe their presentation ideas to the class, and they are discussed. Ideas that do not seem likely to work are revised. At the end of the hour students turn in their written presentation descriptions.

Note:

• A period of time (at *least* one week) should be provided between Hours 3 and 4 for students to prepare their displays and presentations. Students should inform you before Hour 4 if they need presentation aids such as podiums, extension cords, pointers, projectors, tables, or chalkboard space.

HOUR 4: Dress rehearsal: students work on their presentations individually in different areas of the room, gymnasium, auditorium, or wherever the science fair/open house is to be held.

HOUR 5: Students make their presentations every 15 or 20 minutes to small groups of parents who move from one presentation to another.

General Notes About This Project:

• The presentation ideas listed under Hour 2 call for students to make oral presentations. There are many other options, some of which are provided here:

a. A visual display of posters, models, charts, graphs, and reports

b. A "hands-on" activity where guests follow simple instructions to *see* a scientific principle demonstrated as they conduct their own experiment

c. A slide show

d. A video program

e. A computer program

• It is a good idea to establish an "information bureau," consisting of three or four students, to send special invitations for the science fair/open house, and to develop a directory that tells where each student in the class has his or her display. The directory can also give the times for presentations during the evening.

• An open house is an excellent opportunity to transform projects that were taught at a basic level of independence into advanced projects. It may require little independence or use of higher level thinking skills to learn about pulleys, for example, but to *teach* others about what has been learned is a different matter. Planning and presenting a lesson about pulleys is a highly individualized undertaking that requires higher level thinking.

- The open house/science fair serves an important and worthwhile purpose for both students and parents. Students are given an opportunity to make presentations and gain recognition for their learning efforts, and parents get to see their children actively demonstrate a solid grasp of science concepts learned throughout the year. In addition, parents will take pride in their children's ability to learn independently.

Appendix

Teacher's Introduction to the Student Research Guide

Many of the projects in *Learning on Your Own!* require students to conduct research. Few children, however, possess the necessary skills to successfully complete this type of project. The research guide is designed to help them learn and practice some basic skills: how to locate, record, organize, and present information about topics they study. Even though the handouts in the guide are detailed, students will need some guidance and instructional support from you as they undertake their first research projects.

This Teacher's Introduction to the guide contains three forms that can be used to evaluate how well students do on (1) notecards, (2) posters, and (3) oral presentations. These are designed as optional evaluations that may be used with many projects in the book, regardless of subject or topic area.

Teaching a lesson on "how to use the library" is one type of instructional support you should give students to prepare them for research. Therefore, a "Typical Library Quiz" is also included in this Teacher's Introduction for you to give students after they have become acquainted with the library.

The Student Research Guide will be most useful to students if you spend some time explaining the following topics to your class.

1. *Library skills:* Use an example of a real library to explain how books and periodicals are categorized and where they are stored. Whatever library is most likely to be used by students should serve as a model. Cover these things in a library skills unit:

   a. The card catalog (handout provided)
   b. How to find a book from its call number (handout provided)
   c. The *Reader's Guide to Periodical Literature* and other periodical guides (handout provided)
   d. How to ask questions and use librarians as helpful resources
   e. Other kinds of information and services offered by libraries

2. *Notecards and bibliographies:* Provide a variety of examples of properly made notecards and bibliographies for students to use as models. (Reference handout provided.) Explain how to use a numbering system to cross-reference a set of notecards with a bibliography. Spend enough time on bibliographies to ensure

that students know how to write them for the most common sources (books, magazines, encyclopedias, and newspapers).

3. *The Readers' Guide to Periodical Literature*: You can teach students how to use this valuable resource before they ever go to a library. Contact the librarian and ask for old monthly *RGPL* discards. Collect them until you have at least one for every student in the room. During your library skills unit pass the guides out and write ten topics on the board, for example:

   a. The president of the United States
   b. The automobile industry
   c. Basketball (or football, baseball, hockey, and so on)
   d. Ballet
   e. Acid rain
   f. Poland
   g. Israel
   h. Martin Luther King, Jr.
   i. Satellites
   j. Agriculture

   Tell students to choose five topics, find at least one article about each and properly record the title of the article, the author of the article, the name of the magazine, its volume number, the pages on which it can be found, and the date it was published.

4. *Common sources of information:* Encourage students to make extensive use of encyclopedias, magazines such as *National Geographic, Junior Scholastic, Newsweek* and others that are readily available, textbooks and workbooks, materials from home, and whatever other sources are in the classroom or school library. Always require that adequate information from these common sources be available before allowing a research project to begin.

The Student Research Guide is primarily a series of handouts. You may want to give them to students as a complete booklet or hand them out individually to be used with separate, specific research lessons. The research guide is supplied as an aid to help students tackle projects that require research and independent work. The handouts *supplement* what is being taught in the projects, and they provide excellent reference materials for independent learning.

NOTECARD EVALUATION

Below are ten areas for which your notecards have been evaluated. This breakdown of your final score, which is at the bottom of the sheet, indicates the areas where improvement is needed and where you have done well.

	EXCELLENT (10 pts.)	VERY GOOD (9 pts.)	GOOD (7 pts.)	FAIR (6 pts.)	POOR (4 pts.)	NOT DONE OR INCOMPLETE (0 pts.)
1. Bibliography	____	____	____	____	____	____
2. Reference between notecards and bibliography	____	____	____	____	____	____
3. Headings and subheadings	____	____	____	____	____	____
4. Organizing information onto cards so it can be understood and used later without confusion: numbering system	____	____	____	____	____	____
5. Neatness (If reading or use of the cards is made difficult because of sloppy writing,"POOR" will be checked.)	____	____	____	____	____	____
6. Recording meaningful information (Everything recorded on notecards should relate directly to your topic.)	____	____	____	____	____	____
7. Spelling	____	____	____	____	____	____
8. Accuracy of information	____	____	____	____	____	____
9. Quantity (Did you do as much work as you were supposed to, or should have, to complete the project?)	____	____	____	____	____	____
10. Information properly recorded (Facts must be brief and understandable. It is best to condense information into concise statements. Entire paragraphs should not be copied onto notecards. Direct quotes must be identified.)	____	____	____	____	____	____

FINAL SCORE _____ (100 possible)

COMMENTS _____

_____

Name _____ Date _____

POSTER EVALUATION

Below are ten areas for which your poster has been evaluated. This breakdown of your final score, which is at the bottom of the sheet, indicates the areas where improvement is needed and where you have done well.

	EXCELLENT (10 pts.)	VERY GOOD (9 pts.)	GOOD (7 pts.)	FAIR (6 pts.)	POOR (4 pts.)	NOT DONE OR INCOMPLETE (0 pts.)
1. Facts which your poster teaches (at least twenty)	_____	_____	_____	_____	_____	_____
2. Poster "goes along with" your written report	_____	_____	_____	_____	_____	_____
3. Visual impact: use of color, headings, and lettering	_____	_____	_____	_____	_____	_____
4. Drawings (at least one)	_____	_____	_____	_____	_____	_____
5. Pictures, articles, headlines, quotes, charts, graphs, diagrams, explanations, and so forth	_____	_____	_____	_____	_____	_____
6. Organization of material	_____	_____	_____	_____	_____	_____
7. Neatness	_____	_____	_____	_____	_____	_____
8. Spelling, grammar, writing skills	_____	_____	_____	_____	_____	_____
9. Accurate information	_____	_____	_____	_____	_____	_____
10. Specific topic; proper material (Did you do a good job of presenting your topic?)	_____	_____	_____	_____	_____	_____

FINAL SCORE _____ (100 possible)

COMMENTS _____

_____

_____

Name _____ Date _____

ORAL PRESENTATION EVALUATION

This form shows how your oral presentation has been evaluated. It indicates **areas where improvement is needed** and where you have done well.

Topic _____

I. Presentation (50 points possible)

   A. Eye contact. .... 3 pts. _____

   B. Voice projection. .... 3 pts. _____

   C. Use of the English language. .... 3 pts. _____

   D. Inflection. .... 3 pts. _____

   E. Articulation. .... 3 pts. _____

   F. Posture. .... 3 pts. _____

   G. Use of hands. .... 3 pts. _____

   H. Appropriate vocabulary. .... 3 pts. _____

   I. Accurate information. .... 10 pts. _____

   J. Information is easy to understand. .... 3 pts. _____

   K. Enough information. .... 3 pts. _____

   L. Information relates to topic. .... 3 pts. _____

   M. Effort. .... 7 pts. _____

                                                 Subtotal _____

II. Visual or Extra Materials (30 points possible)

   A. Information is easy to understand. .... 3 pts. _____

   B. Information relates to the oral report. .... 3 pts. _____

   C. Information is current. .... 3 pts. _____

   D. Information is accurate. .... 3 pts. _____

   E. Enough information. .... 3 pts. _____

   F. Neatness. .... 3 pts. _____

   G. Spelling. .... 3 pts. _____

   H. Artistic effort. .... 3 pts. _____

   I. Research effort. .... 3 pts. _____

   J. Appropriate vocabulary. .... 3 pts. _____

                                                 Subtotal _____

III. Question-Answer Period (20 points possible)

   A. Confidence in knowledge of topic. .... 3 pts. _____

   B. Ability to answer reasonable questions. .... 3 pts. _____

   C. Answers are accurate. .... 3 pts. _____

   D. Student is willing to admit limits of knowledge or understanding such as "I don't know." .... 2 pts. _____

   E. Answers are brief. .... 3 pts. _____

   F. Student exhibits ability to infer or hypothesize an answer from available information. .... 3 pts. _____

   G. Student appears to have put effort into learning about this topic. ... 3 pts. _____

                                                 Subtotal _____

                             TOTAL (100 pts. possible) _____

COMMENTS _____

_____

TYPICAL LIBRARY QUIZ

How well do you know the library? Answer these questions and find out.

1. List four kinds of information you can find on a card in the card catalog:

   a. _____

   b. _____

   c. _____

   d. _____

2. What does "jB" tell you about a book when it precedes the call number?

   _____

3. What does "jR" tell you about a book when it precedes the call number?

   _____

4. Suppose you are writing a report about polar bears. You look up "polar bears" in the card catalog but find only a few sources. What would you look under next?

   _____

5. If you are looking for a book with the call number j598.132/D43, would you find it before or after j598.2/D42?

   _____

6. If you are looking for "G-men" in the card catalog, you may find a card that says "G-men, see U.S. Federal Bureau of Investigation." Where would you look next?

   _____

7. List these call numbers in the order that they would be found on the shelf:

   j973.15   j973.35   j973.3   j973
   Ad32   Ab24   Cy31   Ad55

   a. _____   c. _____

   b. _____   d. _____

8. Books of fiction are shelved alphabetically by _____.

9. Biographies are shelved alphabetically by _____.

10. What do the words or letters on the front of a card catalog drawer tell you? (example: Istanbul—jets)

   _____

11. How long can books be checked out of the library?

   _____

12. If the book you are looking for is not on the shelf, what should you do?

   _____

13. Where would you go to find a listing of all the magazines your library subscribes to? (Circle the correct answer.)

   a. Card catalog   d. Young adults

   b. History and travel   e. *Readers' Guide to Periodical Literature*

   c. Information desk

14. For *current* information, where should you check first?

   a. Encyclopedia   c. Card Catalog

   b. *Readers' Guide to Periodical Literature*   d. Book shelves

        e. Reference shelves

15. Below is an excerpt from the *Readers' Guide to Periodical Literature*. Look it over and then answer the questions:

   The real cost of a car. S. Porter, il Ladies Home J. 99:58 Je '82

   a. What is the title of the article? _____

   b. Who wrote the article? _____

   c. What month and year was the article published? _____

   d. What magazine published the article? _____

   e. In what volume of the magazine was the article published? _____

   f. On what page can the article be found? _____

   g. Where in the library would you be most likely to find this article?

   _____

Student Research Guide

STUDENT RESEARCH GUIDE

Research is the process you go through to find information about a topic that interests you. This guide explains some basic tools needed to find and record information. It gives advice about how to conduct a research project and also provides many suggestions for developing a *presentation* of the topic.

A list of skills that you will use during research projects includes finding resources, choosing topics, writing and notetaking, summarizing, organizing ideas, scanning, planning, and interpreting data. This guide will help in many of these areas.

Of course, the real quality of a project is determined by the personal characteristics you bring to it—things like patience, motivation, accuracy, neatness, humor, persistence, and creativity. There are no handouts in the Student Research Guide that teach these things, but they are perhaps the most essential ingredients of a successful project.

Your Research Guide contains the following handouts:

"Outlining"
"Bibliographies"
"Notecards and Bibliographies"
"Sending for Information"
"The Dewey Decimal Classification System"
"The Card Catalog"
"The Readers' Guide to Periodical Literature"
"Choosing a Subject"
"Audio-Visual and Written Information Guides"

"Where to Go or Write for Information"
"Project Fact Sheet"
"Project Fact Sheet: Example"
"Poster Display Sheet"
"Things to Check Before Giving Your Presentation"
"Visual Aids for the Oral Presentation"
"Things to Remember When Presenting Your Project"
"Daily Log"
"Blank Skills Chart"

© 1987 by The Center for Applied Research in Education, Inc.

OUTLINING

I. Outlining is like classification: it sorts ideas and facts into categories or like-groups. This is an important skill to have when you are conducting a research project because you must organize information before you can use it.

II. Outlining separates main ideas from details in two ways:

A. By symbols
   1. Alternating letters and numbers
   2. Same symbol = same importance

B. By indentation
   1. Indent more with each subheading
   2. Same margin = same importance

I.

II.

  A.

  B.
    1.
    2.
      a)
      b)
        (1)
        (2)

III. It is very important to understand that every item in an outline can be expanded with additional research or new information. The outline below is incomplete, but it shows how to use symbols and indentation to organize facts and ideas into a logical order. When you make an outline, leave plenty of room between lines so additional ideas can be included later. Think of ways to expand this outline:

EXAMPLE: My Autobiography

I. Early years

  A. Birth

    1. Place

    2. Date

    3. Time

    4. Other details

  B. Family

    1. Father

    2. Mother

    3. Brothers and sisters

    4. Other members of the extended family

    5. Other important adults in your life

  C. First home

    1. Location and description

      a) address

      b) type of house

      c) color

      d) trees in yard

        (1) tall maple in back

        (2) two cherries in front

        (3) giant oak in side yard

          (a) rope swing

          (b) tree house

          (c) shade

            i. summer afternoon naps

            ii. lemonade stand three summers ago

    2. Neighborhood

    3. Experiences

II. School years

  A. School or schools attended

    1. School name and description

    2. Favorite teacher(s)

    3. Favorite subject(s)

  B. Significant experiences

    1. Vacations

    2. Births

    3. Deaths

    4. Adventures

    5. Ideas and beliefs

  C. Friends

III. Present

  A. Residence

  B. Family

  C. School

  D. Hobbies and interests

  E. Friends

IV. Future

  A. Education

  B. Career

  C. Personal goals

  D. Vacations—trips

  E. Family

BIBLIOGRAPHIES

A bibliography is a standard method for recording where information came from. It is important to be able to prove that research came from legitimate sources. Use the following forms when recording information for bibliographies:

I. When working with a book:

    A. Author's last name first
    B. Full title underlined
    C. Place of publication
    D. Date of publication
    E. Publisher
    F. Page(s)

NOTECARD FORM:

Galbraith, John K.
The Affluent Society
Boston
1966
Houghton Mifflin
76

STANDARD FORM:

    Galbraith, John K., The Affluent Society. Boston: Houghton Mifflin, 1966; 76.

II. When working with a periodical:

    A. Author's last name first
    B. Full article title in quotes
    C. Name of periodical underlined
    D. Volume number
    E. Date in parentheses
    F. Page(s)

NOTECARD FORM:

Lippmann, Walter
"Cuba and the Nuclear Race"
Atlantic
211
(Feb. 1963)
55–58

STANDARD FORM:

    Lippmann, Walter "Cuba and the Nuclear Race." Atlantic 211 (February 1963): 55–58.

III. When working with newspaper articles:

    A. Author's last name first
    B. Full article title in quotes

    C. Name of paper underlined
    D. Date
    E. Section (some papers are not divided into sections)
    F. Page

NOTECARD FORM:

May, Clifford D.
"Campus Report: Computers In, Typewriters Out"
The New York Times
May 12, 1986

28

STANDARD FORM:

    May, Clifford D. "Campus Report: Computers In, Typewriters Out," The New York Times, May 12, 1986, p. 28.

IV. When working with an encyclopedia:

    A. Author's last name first
    B. Full title of article in quotes
    C. Name of encyclopedia underlined
    D. Date of publication in parentheses
    E. Volume number
    F. Page(s)

NOTECARD FORM:

Clutz, Donald G.
"Television"
Encyclopaedia Britannica
(1963)
21
910

STANDARD FORM:

    Clutz, Donald G. "Television," Encyclopaedia Britannica (1963), 21, 910.

NOTECARDS AND BIBLIOGRAPHIES

Notecards are used to record and collect information. Bibliography cards are used to tell where the information came from. Once information is gathered about a topic, notecards become the main tool for writing a report. Since each notecard contains a separate idea, you can arrange and rearrange these ideas into an order that becomes an outline for your report. If more information is needed about a particular fact, or, if something needs to be clarified, bibliography cards will tell which source to go to.

Each card should be numbered. It is *very* important that each notecard have a bibliography card number to tell where each fact came from. For example, if you study a unit called "Ecology" in science class, you could do a project about air pollution. Suppose you found information about air pollution in a book titled *Environmental Pollution*—you would make one bibliography card for this source, regardless of how many facts you obtained from it. If this book was the fifth source you used, the bibliography card for it would be numbered "5" in the upper right.

Now, suppose that the chapter on air pollution has four facts, or pieces of information, that you want to use. Make four notecards, each with a unit or course title at the top ("Ecology") and the topic being studied on the next line ("air pollution"). Number these cards in the upper right-hand corner, continuing the numbers from the last card of your fourth source. In other words, if you have 17 notecards from your first four sources, the next card you make will be number 18.

Next, tell where you found the information on each notecard. Do this by writing "bibliography card #5" at the bottom right of each of these four notecards. This clearly shows that you have to look at bibliography card number five to find out where the information came from.

Remember to put only one important fact on each notecard. Don't copy long passages from sources onto notecards; condense information into easily stated facts. If a quote is included in your report, however, it *should* be recorded word for word. Also, if you record your bibliography on notebook paper instead of notecards, each source must still be numbered.

Here is a sample notecard:

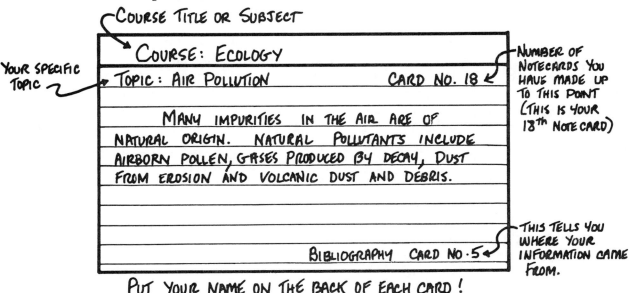

If you are required to record your bibliography on notecards, here is a sample bibliography card:

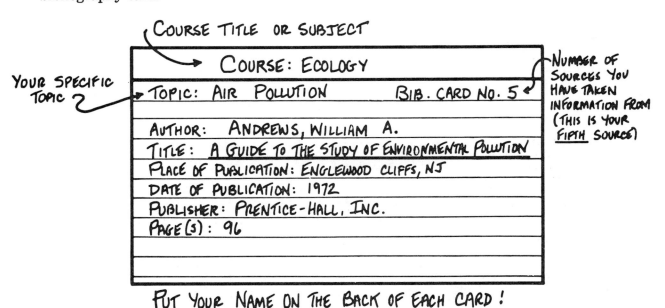

SENDING FOR INFORMATION

There are times when sending a letter is the best way to obtain information about a research topic. Unfortunately, many people write letters hastily. They don't take time to explain themselves clearly or else they come across sounding unprofessional and insincere. You should learn how to write a good letter so that, when confronted with a difficult project, you can get help from others. Study the outline below. It explains why letter writing is a useful research skill and what components should be included in the letters you write. Examples of two letter styles are provided.

I. Reasons for sending a letter:

    A. To obtain up-to-date information.

    B. To make contact with experts or specific organizations.

    C. To get specialized or technical information.

    D. To ask for opinions and advice.

    E. To ask for suggestions of other places to look for information about the topic.

    F. To ask for free materials.

    G. To send special questions to authorities in the field you are studying.

II. Parts of a letter

A. Heading:	*Your* return address at the top of the letter, and the date right below your address.
B. Inside address:	The address of the person or organization to whom you are sending the letter.
C. Salutation:	Begin your letter with a salutation to the person you are sending it to: Dear Mr. Wilson; Dear Miss Goode; Dear Mrs. Smith; Dear Ms. Jones; Dear Sir.
D. Body:	Introduce yourself, explain your project, and ask for whatever assistance you are seeking. Be concise and clear in your writing; don't make someone guess what you want.
E. Complimentary close:	Show your respect by thanking the person to whom you have sent your letter for whatever help he or she can provide. Your letter might end like this:

"…I appreciate any advice or information you can offer to help me with my project.
Thank you."

Sincerely,

John Jones

F. Signature	Sign your name at the bottom of the letter, beneath the complimentary close.

EXAMPLE OF THE "BLOCK LETTER" STYLE

John Jones
1532 Hill Street
Bridgeton, TX 75588

March 16, 19XX

Dr. David Adamson
Entomological Society
113 Geneva Road
Fair Ridge, OH 45289

Dear Dr. Adamson:

I am an eighth-grade student at Bridgeton Middle School, and we are doing a science project on insects. I am studying the praying mantis, and I have three questions that I can't find answers to from my research. I thought maybe you could help me.

I have enclosed a self-addressed, stamped envelope for your convenience. Here are my questions:

1. By what other names are praying mantises known?
2. How many species are there?
3. Can young praying mantises fly?

I appreciate any information you can provide about these questions. Thank you.

Sincerely,

John Jones

John Jones

EXAMPLE OF A "MODIFIED BLOCK LETTER" STYLE

Dr. David Adamson
Entomological Society
113 Geneva Road
Fair Ridge, OH 45289

March 22, 19XX

John Jones
1532 Hill Street
Bridgeton, TX 75588

Dear John,

I received your letter of March 16, and I am glad to help you. Here are my answers to your questions:

1. The praying mantis is also known by these names: rearhorse, mule killer, devil's horse, and soothsayer.
2. There are 20 species of praying mantis. The European mantis is well established in the eastern U.S., and the Chinese mantis has also established itself in the eastern states.
3. One female lays up to 1,000 eggs in the fall, which hatch in May or June. The young cannot fly; they grow slowly, acquiring wings and maturity in August. When mature, four well-developed wings allow slow, extended flight.

I hope this information helps you in your research work. By the way, thank you for enclosing a stamped envelope—I appreciate that. If I can be of further assistance, please let me know.

Sincerely,

David Adamson, M.D.

Dr. David Adamson

Name _____ Date _____

DEWEY DECIMAL CLASSIFICATION SYSTEM

1. The Dewey Decimal Classification System arranges all knowledge into ten "classes" numbered 0 through 9. Libraries use this system to assign a "call number" to every book in the building. A call number is simply an identification number that tells where a book is located in the library.

   (000) 0—Generalities
   (100) 1—Philosophy and related disciplines
   (200) 2—Religion
   (300) 3—The social sciences
   (400) 4—Language
   (500) 5—Pure sciences
   (600) 6—Technology (applied sciences)
   (700) 7—The arts
   (800) 8—Literature and rhetoric
   (900) 9—General geography, history,
   　　　　and so forth

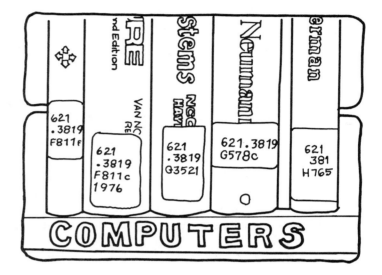

2. Each *class*, with the use of a three-digit number, is divided into ten subclasses (divisions) with the first division (600–609) set aside for the general works on the entire class. For example, 600–649:

   600–609 is given over to *general works* on the applied sciences
   610–619 to the medical sciences
   620–629 to engineering and applied operations
   630–639 to agriculture and agricultural industries
   640–649 to domestic arts and sciences

3. Each division is separated into ten subclasses or *"sections"* with the first *"section"* (630) devoted to the general works on the entire *division*. For example:

   630 is assigned to agriculture and agricultural industries in general
   631 to farming activities
   632 to plant diseases and pests and their control
   633 to production of field crops
   636 to livestock and domestic animals, etc.

4. Further subdividing is made by following the three-digit number with a decimal point and as many more digits as is necessary. For example, 631 farming is divided into

   631.2 for farm structures
   631.3 for farm tools, machinery, appliances
   631.5 for crop production

5. In summary, every book in a library is assigned a call number based upon the Dewey Decimal Classification System. All library books are stored on shelves according to their numbers, making them easy to find.

6. To locate a book in the library follow these steps:

   a. Use the card catalog to find the call number of a book in which you are interested. Books are cataloged by author, title, and subject.
   b. Record the call number, usually recorded in the upper left-hand corner of the card. If your library uses a computerized catalog system, ask a librarian for assistance in locating the call number.
   c. Refer to the first three numbers of the call number to determine in which section of the library your book can be found.
   d. Once you have found this section of the library, use the rest of the call number to locate the book on the shelf.

THE CARD CATALOG

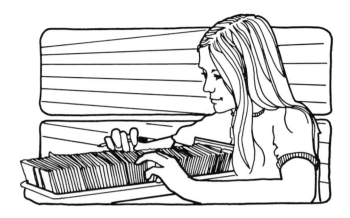

The card catalog is usually the first place you would go to look for a book in the library. The cards in the card catalog are arranged alphabetically by subject, author, and title. The card below is a "subject" card, filed under "inventors." The same book could be found if you looked under "Manchester, Harland Frank" (along with any other books Mr. Manchester has written) or *Trailblazers of Technology* (the title of the book).

Once you find the card that best fits your needs, the most important piece of information is the "call number" in the upper left-hand corner. This number tells you where to find the book in the library. In trying to decide which book to look up, you may refer to various pieces of information found on every catalog card. This information includes

1. Call number
2. Subject
3. Author
4. Author's birth date
5. Title
6. Brief description
7. Illustrator (if there is only one)

8. Location of publisher (city)
9. Publisher
10. Date of publication
11. Number of pages
12. Whether or not the book is illustrated
13. Size of the book

Here is a sample card:

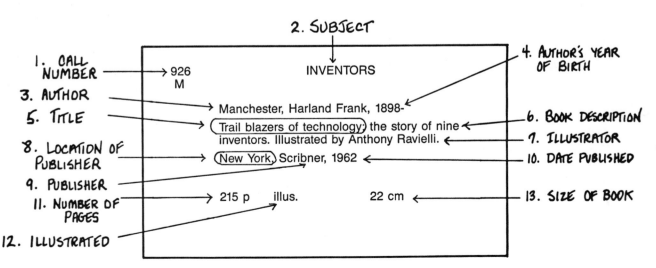

READERS' GUIDE TO PERIODICAL LITERATURE

The *Readers' Guide to Periodical Literature* is an extremely useful tool. You can find magazine articles about current topics from as recent as one or two months ago. You can also find articles that were written fifteen, twenty, or fifty years ago. Subject and author headings are arranged alphabetically in the *Guide*. Articles are arranged alphabetically under each heading.

When a promising reference is found, first determine how to locate the magazine that published the article. Does the library subscribe to the magazine? Is it a current issue? (Usually issues for the past twelve months will be available in the periodical reading section of the library.) Are old issues recorded on microfilm, or are they bound and placed in a special area of the library? When you find an article you want to read, record the following on a piece of paper. Then you or a librarian, if necessary, can locate the magazine from its file area.

The elements in each entry are:

1. Title of the article
2. Author's name
3. Name of the magazine
4. Volume number
5. Pages on which the article can be found
6. Date of publication

Suppose you are studying atomic power; specifically, you want to find out about the costs of building and operating atomic power plants. By looking through a *Readers' Guide*, you will find many articles published about atomic power. From the example provided you can see that "atomic bombs" is at the top of the list of atomic topics, followed by atomic energy, atomic energy industry, atomic facilities, atomic fuels, atomic power, atomic power industry, atomic power plants, and atomic research. There, under "Atomic power plants—Economic aspects," is a collection of five articles that could be useful to you. Look at the fourth one:

Whoops! A $2 billion blunder. C.P. Alexander. il *Time* 122:50-2 Ag 8 '83

1. Title: "Whoops! A $2 Billion Blunder"
2. Author: C.P. Alexander
3. il: this article is illustrated with photographs or drawings
4. Magazine: *Time*
5. Volume: 122
6. Page(s):   50-52
7. Date: August 8, 1983

© 1987 by The Center for Applied Research in Education, Inc.

READERS' GUIDE TO PERIODICAL LITERATURE (continued)

The *Readers' Guide to Periodical Literature* makes use of abbreviations for months of the year, magazine names, and other pieces of important information. For example, "Bet Hom & Gard" is *Better Homes and Gardens* and "bi-m" means a magazine is published bimonthly. Be sure to refer to the first few pages of the *Readers' Guide* for a complete list of all the abbreviations used.

The following is a section from the *Readers' Guide to Periodical Literature*:

Atomic bombs
> **History**
> *See also*
> Hiroshima (Japan)
> **Physiological effects**
> *See* Radiation—Physiological effects
> **Testing**
> *See* Atomic weapons—Testing

Atomic energy *See* Atomic power

Atomic energy industry *See* Atomic power industry

Atomic facilities *See* Nuclear facilities

Atomic fuels *See* Nuclear fuels

Atomic power
> *See also*
> Anti-nuclear movement
> Nuclear fuels
> **Economic aspects**
> *See also*
> Atomic power industry
> **Laws and regulations**
> *See also*
> Radioactive waste disposal—Laws and regulations
> Mixed rulings on nuclear power [Supreme Court decisions] R. Sandler. *Environment* 25:2-3 Jl/Ag '83
> **Germany (West)**
> *See also*
> Anti-nuclear movement—Germany (West)

Atomic power industry
> *See also*
> Computers—Atomic power industry use
> Reactor fuel reprocessing
> Washington Public Power Supply System
> **Export-import trade**
> Firing spotlights plutonium exports [R. Hesketh's claim that plutonium produced in Great Britain's civilian reactors has been used in U.S. weapons manufacture] D. Dickson. *Science* 221:245 Jl 15 '83
> **Laws and regulations**
> *See* Atomic power—Laws and regulations
> **Public relations**
> Atom and Eve [nuclear acceptance campaign geared to women] L. Nelson. il *Progressive* 47:32-4 Jl '83
> **United States**
> *See* Atomic power industry

Atomic power plants
> **Economic aspects**
> The bankruptcy of public power [Washington Public Power Supply System debacle] *Natl Rev* 35:982-3 Ag 19 '83
> Money meltdown [Washington Public Power Supply System default] S. Ridley. *New Repub* 189:11-13 Ag 29 '83
> When billions in bonds go bust [default of Washington Public Power Supply System] *U S News World Rep* 95:7 Ag 8 '83
> Whoops! A $2 billion blunder. C. P. Alexander. il *Time* 122:50-2 Ag 8 '83
> The Whoops bubble bursts. H. Anderson. il *Newsweek* 102:61-2 Ag 8 '83
> **Laws and regulations**
> *See* Atomic power—Laws and regulations
> **Safety devices and measures**
> Computers to supervise nuke plants. *Sci Dig* 91:27 Jl '83

Atomic research
> Pseudo-QCD [discussion of January 1983 article, A look at the future of particle physics] B. G. Levi. *Phys Today* 36:98+ Jl '83

CHOOSING A SUBJECT

The first step in any research project is choosing something to study. This requires some thought and decision making. This handout provides several guidelines that will help you select a subject.

1. Choose a subject that you are already interested in or that you would like to know more about.

2. Choose a subject that will meet the needs or requirements as outlined by the teacher:
   a. Listen for suggestions from the teacher.
   b. Be alert to ideas that come from class discussion.
   c. Talk to friends and parents about things you can study and learn.

3. A good rule by the Roman poet Horace: "Choose a subject, ye who write, suited to your strength." This means pick a subject you can understand, not one in which you will become bogged down, lost, or disinterested.

4. The encyclopedia should serve as a tool for choosing the right subject and narrowing it down so you can handle it:
   a. It gives the general areas of the subject.
   b. It identifies specific topics related to your subject.
   c. It is written simply enough to understand without hours of study.

5. Before you commit yourself to a subject, check to make sure there is some information available. There is nothing more frustrating than starting a project that cannot be finished because there are no books, magazines, filmstrips, newspapers, journals, experts, or even libraries that have enough information.

6. Once you have chosen a subject, write down a series of questions to which you want to find answers. Write as many as you can think of. These questions will help direct your research.

AUDIO-VISUAL AND WRITTEN INFORMATION GUIDES

DIRECTIONS: The following list shows some of the places where information can be found. When you begin your first project, go down column one and put a check mark in the box next to each place you *might* be able to find information. When you *do* find information, fill in the appropriate box on the chart with your pencil. Do this for your first five research projects.

	PROJECT NUMBER				
	1	2	3	4	5
Almanacs					
Atlases					
Bibliographies					
Biographies					
Charts and graphs					
Dictionaries					
Encyclopedias					
Films					
Filmstrips					
Historical stories					
Indexes to free material					
Letters					
Library card catalog					
Magazines					
Maps					
Microfilm					
Newspapers					
Pictures					
Readers' Guide to Periodical Literature					
Records					
Tapes					
Textbooks					
Vertical files					
Other: _____					
_____					

Name _____ Date _____

WHERE TO GO OR WRITE FOR INFORMATION

DIRECTIONS: Before you start your project, put a check mark in the box next to each place you could go or write to get information. When you *do* get information, fill in the appropriate box.

	PROJECT NUMBER				
	1	2	3	4	5
Chambers of Commerce					
Churches					
City officials					
Companies					
Embassies					
Experts					
Factories					
Federal agencies					
Historical societies					
Hobbyists					
Librarians					
Libraries					
Ministers					
Museums					
Newspaper office/employee					
Organizations (club, societies)					
Professionals					
Research laboratories					
State agencies					
Teachers					
Travel agencies					
Universities					
Zoos					
Friends					
Home (books, magazines, etc.)					
Other: _____					
_____					

Name _____ Date _____

PROJECT FACT SHEET

One of the most difficult parts of any project is getting started. Use the "Project Fact Sheet" to begin recording information that will be included in a presentation or report. A sample of a completed "Project Fact Sheet" is shown on the next page.

My topic is _____

and these are the facts I am going to teach the rest of the class:

1. _____
2. _____
3. _____
4. _____
5. _____
6. _____
7. _____
8. _____
9. _____
10. _____
11. _____
12. _____
13. _____
14. _____
15. _____
16. _____
17. _____
18. _____
19. _____
20. _____

Name _____ Date _____

PROJECT FACT SHEET: Example

This sample fact sheet about humpback whales shows how to write out information that is to be included in a presentation.

My topic is <u>Humpback Whales,</u> and these are the facts I am going to teach the rest of the class:

1. Humpback whales spend six months in the South Pacific.
2. Humpback whales sing a strange song that seems to be some sort of communication.
3. Humpback whales sing only when they are in the South Pacific.
4. Humpback whales do not eat when they are in the South Pacific.
5. Humpback whales travel to an arctic Alaskan bay to feed.
6. A humpback whale has a brain that is five times larger than a human brain.
7. The invention of the explosive harpoon gun and the steam engine made full-scale hunting of the humpback whale possible.
8. Humpback whales show great devotion to one another; this is best displayed by the relationship between a mother and her young.
9. A young whale is called a "calf."
10. The humpback whale eats krill, which makes it a carnivorous mammal.
11. (This list is extended to whatever the project outline requires.)

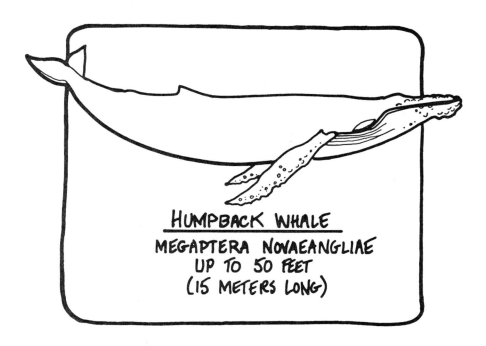

HUMPBACK WHALE
MEGAPTERA NOVAEANGLIAE
UP TO 50 FEET
(15 METERS LONG)

POSTER DISPLAY SHEET

Use the guidelines on this handout if you are required to make a poster for a research project.

1. Present or "teach" at least twenty facts about your topic on the poster. These facts should be recorded on notecards.

2. The poster should be made to go with the written report so that they can be used together when you make a presentation.

3. Include at least one of your own drawings on it.

4. The poster can also have other pictures, magazine articles, newspaper headlines, quotes from books, charts, graphs, illustrations, explanations, diagrams, captions, and so forth.

5. Organize all of the material on the poster so that it is easy to understand. This is very important when making a top-quality poster. Give your poster visual impact by using colorful designs, bold headings, and a catchy title.

6. Writing must be neat! Use parallel guidelines and pencil words in lightly before going over them with marker.

7. Check spelling, grammar, capitalization, punctuation, and sentences to be sure they are correct.

8. Every bit of information you use must be accurate. *Do not make anything up!*

9. Your poster should be about a very specific topic. Don't throw everything you can find onto it. Be selective and use only material that contributes favorably to the project.

10. OPTIONAL: Write five questions that can be answered by studying your poster. These questions should be attached to the poster.

Name _____ Date _____

THINGS TO CHECK BEFORE GIVING YOUR PRESENTATION

DIRECTIONS: After practicing your presentation at home one time, write "yes" or "no" in the boxes below to help determine which areas need more work. The purpose of this checklist is to help put *quality* into your presentation. Use it wisely and be honest. If something needs more time and effort, be willing to admit it and work to improve what you have done.

	PROJECT NUMBER				
	1	2	3	4	5
Have I done enough research?					
Is everything spelled correctly?					
Did I use neat handwriting?					
Is everything in my visual display labeled?					
Do all my pictures have captions?					
Is my visual display neat and attractive?					
Did I use colors in a pleasing way?					
Did I do my best artwork?					
Does my oral report need more practice?					
Do I know all the words in my report?					
Is it easy to understand what I have written?					
Is my report informative?					
Is my visual display informative?					
Do I understand the information I will present?					
Did I choose interesting and different presentation methods?					
Have I decided how I will display my visual materials during my presentation?					
Am I ready to answer questions about my subject?					
Did I follow the project directions or outline?					
Does my presentation stick to my subject?					
Is this my best work?					

Name _____ Date _____

VISUAL AIDS FOR THE ORAL PRESENTATION

DIRECTIONS: Making your report interesting is very important. Besides hearing what you have to say, the audience likes to see examples of what you've done. There are many ways to use visual aids during a presentation. This list provides some suggestions. First, check the items that you think you *could* use. Later, fill in the ones you actually *did* use.

	PROJECT NUMBER				
	1	2	3	4	5
Chalkboard					
Charts					
Clippings					
Diagrams					
Dioramas					
Film (slides)					
Filmstrips					
Guest speakers					
Magazines					
Maps					
Models					
Murals					
Opaque projector					
Overhead projector					
Pictures					
Posters					
Records					
Tape recorder					
Other: _____ _____					

© 1987 by The Center for Applied Research in Education, Inc.

When speaking to a group you must always be aware of these things:

1. Voice projection
2. Eye contact
3. Inflection

4. Proper grammar
5. Hand control
6. Posture

THINGS TO REMEMBER
WHEN PRESENTING YOUR PROJECT

Try to remember these rules when you are speaking before the group. Underline the ones you need to improve. On the lines at the bottom of this sheet, write any other rules and notes you feel you need as reminders.

1. Speak in complete sentences.
2. Use any new vocabulary words you may have learned, but be sure you can pronounce them and that you know what they mean.
3. Speak with a clear voice so that everyone can hear.
4. Look at your audience and speak to its members.
5. Stand aside when you are pointing out pictures, maps, charts, drawings, or diagrams.
6. Do not read long passages from your notes.
7. Know your material so that you sound like an informed person.
8. Be as calm as possible. Try to show that you have confidence in your work.
9. Do not chew gum when presenting.
10. Be ready to tell where you got your information.
11. Explain what your visual display shows, but don't read everything that is on it to your audience. Let the audience read it later.
12. Ask for questions from the class.
13. Be willing to admit that you don't know an answer if you really don't know.
14. Never make up an answer. You are expected to give only accurate information.
15. When your project is due to be presented, have it ready in final form—and on time! Do not come to class with empty hands and a list of excuses.

16. _____

_____

17. _____

_____

18. _____

_____

NOTES: _____

_____

_____

HOW TO USE THE DAILY LOG

Directions:

One of the most important requirements of an independent worker is an accurate record of each day's accomplishments. This is especially important for students who are just learning how to do research projects on their own. A Daily Log is helpful because every step of the project is recorded. This allows the teacher to check your progress without watching you work. The more conscientious you are about keeping a detailed, accurate log, the more likely you are to earn the right to become involved in even more independent projects.

To use the log on the next page, simply fill in the date and the time you started working on your project. Describe what you did as accurately as possible and record what was accomplished. Record the time when you are finished.

For example:

Oct. 14   10:45–11:25   Looked in 3 magazines for info. about earthquakes. Recorded facts on 10 notecards. Found 2 poster ideas.

DAILY LOG

Name: _____

Project Title: _____

Date Due: _____ Date Begun: _____ Date Completed: _____

DATE	TIME BEGUN	TIME ENDED	DESCRIPTION OF WORK

SKILLS CHART: SCIENCE

		RESEARCH									WRITING						PLANNING				
# *Prerequisite Skills* Students must have command of these skills.																					
X *Primary Skills* Students will learn to use these skills; they are necessary to the project.		PREPARING BIBLIOGRAPHIES	COLLECTING DATA	INTERVIEWING	WRITING LETTERS	LIBRARY SKILLS	LISTENING	MAKING NOTECARDS	OBSERVING	SUMMARIZING	GRAMMAR	HANDWRITING	NEATNESS	PARAGRAPHS	SENTENCES	SPELLING	GROUP PLANNING	ORGANIZING	OUTLINING	SETTING OBJECTIVES	SELECTING TOPICS
0 *Secondary Skills* These skills may play an important role in certain cases.																					
* *Optional Skills* These skills may be emphasized but are not required.																					

SKILLS CHART: SCIENCE

	PROBLEM SOLVING						SELF-DISCIPLINE										SELF-EVALUATION				PRESENTATION								
	BASIC MATHEMATICS SKILLS	DIVERGE-CONVERGE-EVALUATE	FOLLOWING & CHANGING PLANS	IDENTIFYING PROBLEMS	MEETING DEADLINES	WORKING w/LIMITED RESOURCES	ACCEPTING RESPONSIBILITY	CONCENTRATION	CONTROLLING BEHAVIOR	FOLLOWING PROJECT OUTLINES	INDIVIDUALIZED STUDY HABITS	PERSISTENCE	SHARING SPACE	TAKING CARE OF MATERIALS	TIME MANAGEMENT	WORKING WITH OTHERS	PERSONAL MOTIVATION	SELF-AWARENESS	SENSE OF "QUALITY"	SETTING PERSONAL GOALS	CREATIVE EXPRESSION	CREATING STRATEGIES	DIORAMA & MODEL BUILDING	DRAWING/SKETCHING/GRAPHING	POSTER MAKING	PUBLIC SPEAKING	SELF-CONFIDENCE	TEACHING OTHERS	WRITING